HUTCHINSON

Trends in Science

CHEMISTRY

HUTCHINSON
Trends in Science
CHEMISTRY

Overview by
Dr Keith B Hutton

FITZROY DEARBORN PUBLISHERS
CHICAGO • LONDON

Copyright © Helicon Publishing 2001

All rights reserved

Published in the United States of America by
Fitzroy Dearborn Publishers
919 North Michigan Avenue
Chicago, Illinois 60611
OR
Fitzroy Dearborn Publishers
310 Regent Street
London W1B 3AX

First published in the USA and in the UK 2001

ISBN: 1–57958–359–8

Library of Congress and British Library Cataloguing in Publication Data are available

Typeset by
Florence Production Ltd, Stoodleigh, Devon
Printed and bound in Great Britain by
Clays Ltd, Bungay, Suffolk

Editorial Director
Hilary McGlynn

Managing Editor
Elena Softley

Project Editor
Heather Slade

Editor
Catherine Thompson

Technical Editor
Rachel Margolis

Production Manager
John Normansell

Production Controller
Stacey Penny

Picture Researcher
Sophie Evans

Contents

Preface

The definition of chemistry given in the dictionary is 'the science of the elements and the laws of their combination and behaviour under various conditions'. This rather dry definition does not do justice to the thousands of individuals who developed chemistry into one of the most influential sciences that has helped to shape the modern world.

The scope of the science is vast and began in prehistoric times when humans first produced heat from carbon, the discovery of fire. Philosophers such as Plato and Aristotle in Ancient Greece discussed the theory of chemistry centuries before the rise of the Roman Empire. The first practitioners of what would become chemistry were the secretive and solitary alchemists who searched for a method to transmute base metals into gold.

The discoveries of the medieval period paved the way for the birth of modern chemistry in the 17th and 18th centuries. Great pioneering chemists such as Boyle and Lavoisier swept away the superstitions and inconsistencies of the past and reinvented chemistry using firm scientific principles. For the first time, chemical reactions between substances could be accurately predicted and controlled, which in turn led to the mass production of the chemicals that helped to make the Industrial Revolution a reality.

By the end of the 19th century, natural products such as silk, dyes, and wood were gradually being replaced by synthetic equivalents made by chemists. The 20th century would see advances in the science that would affect virtually every person on Earth. The best and the worst of what chemists had to offer was still waiting to be discovered.

This crucial period in the history of chemistry is the subject of this publication. The major achievements of the 20th century are summarized in an extensive chronology covering the whole scope of the science, from the development of the petrochemical industry to the discovery of vitamins, antibiotics, and electrically conducting plastics. And an insight into the lives of those scientists who have contributed to the development of chemistry during this century can be gained from reading their biographies and what they have achieved.

Every science has its own specialized language to describe what it does and chemistry is no exception. A selected glossary of definitions and terms used by chemists, as well as a detailed bibliography of relevant publications, are provided to help give the reader an appreciation of how chemists describe their world.

Those institutions, organizations, and companies currently involved in the science, best illustrate the present state of chemistry. *Trends in Science:*

Chemistry provides descriptions and contact information for the major organizations carrying on the work of the chemists of the past.

An accurate perspective of what impact chemistry has had on human society can only be gained if a step back is taken to review those innovations that have had a global impact. *Trends* provides a detailed overview of the more significant advances in chemistry, how they developed and how they continue to influence our attitudes towards the science. Darker aspects, such as chemical weapons and environmental pollution, counterbalance triumphs such as the development of drugs and plastics. However, as well as the past, the overview also looks to the future and the role that the chemist may play in the 21st century.

Dr Keith B Hutton
Clarendon Laboratory
University of Oxford
January 2001

Part One

1 Overview

Introduction – chemistry at the start of the 20th century

By 1900, chemistry had come a long way from its origins in medieval times with the alchemists. The understanding of the elements and chemical reactions had become a distinct science separate from natural philosophy, or physics as it is now known. The existence of the atom was still in dispute, although most scientists accepted it as a useful concept. The first subatomic particle, the electron, had been discovered, but scientists did not fully understand what they had found.

Many new elements had been discovered in the 19th century and a system for their classification had been established. It was called the periodic table and was constructed by placing the elements in order of increasing atomic weight, the assumption being that this was the property that characterized an element. The elements were arranged in columns such that similar chemical properties occurred at fixed intervals or periods. Sometimes this arrangement could not be made to work and so gaps were left in the table, anticipating the discoveries of new elements. The search for the missing elements had led to the discovery of new radioactive members of the table. However, isolating these elements for study had proven to be difficult and so little was known about them.

Also in the 19th century, chemical production had become an integral part of industrialized society. The majority of the technology had evolved in Germany where the availability of huge coal reserves had influenced the development of the industry. There, chemists had concentrated on the conversion of coal to produce chemicals. A disadvantage with this approach was that the industrial processing of crude oil, or petroleum, was totally neglected in Europe. One class of compounds that were not produced by coal-conversion were the unsaturated hydrocarbons from which synthetic plastics are produced. The development of plastics would have to wait for the rise of the mighty US petrochemical industry.

In 1828, German chemist Friedrich Wöhler (1800–1882) had produced the organic chemical urea using only inorganic starting materials, something thought impossible at the time. Synthesis, the artificial production of a substance from its constituents, had been restricted to inorganic compounds, but now new doors had been opened to the chemist. Progress was hampered by having no reliable techniques to examine how a substance was chemically constructed, leaving the synthetic chemist to stumble blindly through the jungle of chemical preparations, occasionally stumbling onto the correct path. Progress relied on individual breakthroughs rather than systematic development. Consequently, by the start of the 20th century, synthetic dyes were the only organic compound in full commercial production.

Chemistry was on the brink of a scientific and industrial explosion of development, which would see the discovery of wonders and the creation of nightmares to make society pause to consider whether it had chosen the right path.

The atom – physicists lend a hand

At the beginning of the 20th century, the concept that matter was composed of indivisible particles known as atoms was universally accepted. This approach had been building momentum since the beginning of the 19th century when English chemist John Dalton (1766–1844) proposed his 'atomic theory' to explain how compounds are formed. However, although the theory was accepted and widely used to explain chemical behaviour, there was no direct evidence that atoms and molecules existed and some prominent scientists still maintained that they were only convenient abstractions.

Then, in 1908, French physicist Jean Perrin (1870–1942) found the proof that molecules were real. It had been suspected since the middle of the 19th century that Brownian motion, the random movement of particles suspended in water, was caused by the particles colliding with water molecules. German-born US physicist Albert Einstein (1879–1955) assumed this to be the case and published his theoretical analysis of the phenomenon in 1905. In this study, he derived a formula which could be used to calculate the size of the water molecule. Perrin added a dye of known particle size to a cylinder of water and then studied the manner in which it settled. He noticed that the downward passage of the particles was being opposed by Brownian movement. Since the only force acting on the dye was gravity, Perrin could calculate the force being applied by the water to oppose the downward motion. This allowed him to use Einstein's formula to calculate the size of a water molecule. The results of his calculations convinced the sceptics that molecules did exist.

The discovery of the negatively charged electron, by English physicist J J Thomson (1856–1940) in 1897, had proved that the atom was divisible, but little else was known about atomic structure. Thomson proposed a 'plum pudding' model of the atom, where the majority of the mass was the positive sphere of the pudding, which had the negatively charged electrons imbedded like fruit inside it. This idea was generally accepted until New Zealand-born British physicist Ernest Rutherford (1871–1937) announced his concept of the nuclear atom in 1911. The research group he lead had been studying the effects of firing alpha particles at thin foils of gold and platinum using a photographic plate placed behind them as a target. Although most of the particles had passed straight through, he observed an unexpected scattering around the central spot. By positioning the photographic plate it was discovered that some particles had been deflected by large angles and some had even been bounced back along the path of the incoming alpha particle. Rutherford commented that this was the equivalent of shooting a 15-inch shell at a piece of tissue paper and then being hit by a rebound. Rutherford formed the correct conclusion that the alpha particles had struck something very dense in the foil and, since so few collisions had occurred, whatever this was must be very small. He concluded that he had found the nucleus. Rutherford's model of the atom had almost all of its mass located in a small, positively charged core called the nucleus, surrounded by a mist of electrons that occupied most of the atom's space.

Why negatively charged electrons were not drawn into the positive nucleus was explained in 1913 by Danish physicist Niels Bohr (1885–1962). He suggested that electrons were fixed into circular orbits (shells) around the nucleus. Drawing on the 1900 theory of German physicist Max Planck (1858–1947) that radiation was emitted in fixed amounts or 'little packets' called quanta, Bohr argued that no electrons were present in the gaps between the shells of an atom and that the energy difference between each shell was exactly a quantum specific to that element. As only a limited number of electrons were allowed to occupy each shell, this theory resulted in all elements having a unique electronic configuration. Bohr's theory was crucial in explaining certain chemical properties of the elements.

Rutherford continued to study the bombardment of metals by alpha particles. He noticed that in every case, positively charged hydrogen ions were being given off as a by-product of the process. Hydrogen is the simplest element and Rutherford correctly concluded that the hydrogen ion was the positive counterpart of the electron in the atom. In 1920 he published his results and called this fundamental positive particle the proton from the Greek word meaning 'first'.

In 1932, English physicist James Chadwick (1891–1974) discovered neutrons. These electrically neutral particles act as the 'glue' which binds the positively charged protons together to form the nucleus. With their discovery, the structure of the most basic building block in chemistry, the atom, had been found.

The periodic table – the last pieces of the puzzle come together

In 1869, Russian chemist Dmitri Mendeleyev (1834–1907) had arranged the known elements into the periodic table – so called because it showed the periodic recurrence of elements with similar chemical properties and was based on the atomic weight of the element. However, Mendeleyev reasoned that the chemical properties of an element were more important than weight and he showed no hesitation in swapping elements if their properties did not match the rest of the series. More radically, if he could not get the order of the table to work with the known elements, he left gaps for future discoveries.

The search to find the missing elements led to the discovery of new radioactive members of the periodic table. Although radium had been discovered in 1898, the isolation of the element in the form of a pure compound had to wait until 1902. This achievement was a triumph of chemistry for Polish-born French nuclear chemist Marie Curie (1867–1934) and her husband, French physicist Pierre Curie (1859–1906). They had taken four years of painstaking effort to extract 1 g/0.04 oz of pure radium chloride from 8 tonnes of the uranium-rich mineral pitchblende. In 1900 German physicist Friedrich Dorn (1848–1916) had a considerably easier task when he discovered that radium decayed to produce the radioactive gas radon. He simply collected the gas in a glass tube.

Transmutation of elements

It soon became clear that something unusual was happening in pitchblende. Studies of this mineral had led to the discovery of polonium and radium in 1898, actinium in 1899, and protactinium in 1913. It was correctly deduced by Rutherford and his English assistant physical chemist Frederick Soddy (1877–1956), that in the act of producing radiation, the elements were changing into other radioactive elements. The transmutation of elements, long searched for by the alchemists, had been found!

Chemists began looking eagerly for new transformations, but their efforts were rewarded by a bewildering assortment of compounds. It became clear that the reliance on characterizing the elements by their atomic weight led to ambiguous and misleading 'discoveries'. During this period many more 'elements' were claimed to have been found than there was space for in the periodic table and one by one each was identified as being chemically indistinguishable with a known element, but possessing a different atomic weight.

Radioactive isotopes

So what was going on? The evidence suggested that radioactive elements could exist in more than one form. English physical chemist Frederick Soddy (1877–1956) expressed this idea fully in 1913 when he described the various forms of each element as being isotopes, from the Greek for 'the same position'. However, it was still not known if the concept of isotopes was confined to the radioactive elements. The issue was decided with the help of English

Marie Curie and her husband Pierre in their Paris laboratory. They received the Nobel Prize for Physics in 1903 for the discovery of radioactivity. In 1911 Marie Curie became the first person to be awarded the Nobel prize twice, when she was awarded the chemistry prize for her discovery of radium. Some of her notebooks are so radioactive that they still cannot be handled. AEA Technology physicist Francis William Aston (1877–1945) with his invention of the mass spectrograph. This machine separated out isotopes by the virtue of deflections of their ions by a magnetic field. In 1919 J J Thomson used it to show that neon existed in at least two forms. Aston carried on the work to show that several more stable elements were composed of isotopes, notably chlorine, for which feat he was awarded the Nobel Prize for Chemistry in 1922. It is now known that very few elements exist in nature as a single isotope. Following a suggestion by US chemist Truman Kohman in 1947, the nucleus of the atom is now called a nuclide. Isotopes are identified by their atomic mass number, for example neon exists as three isotopes, neon-20, neon-21, and neon-22.

Atomic number

A property for distinguishing between elements other than atomic weight had to be found. It took a physicist, the English physicist Henry Moseley (1887–1915), to find it. In 1913 Moseley discovered that the X-ray spectra of the elements had a deviation that changed regularly through the periodic table. A graph of the square root of the frequency of each radiation against a quantity he called the atomic number of the element, gave a straight line. Atomic number was shown later to correspond to the number of protons in the nucleus of the atom. This quantity is the characteristic feature of an element because all of its isotopes, whatever their mass, have the same atomic number. This fact fixes an element to its place in the periodic table. Moseley published his findings in 1914 and scientists used his discovery to draw up a new periodic table where the elements were arranged in order of their atomic numbers. This is the form in which the periodic table is used today, although atomic number is now called proton number.

Before Moseley, only two elements had been conclusively identified since the turn of the 20th century, radon and lutetium. Lutetium was isolated by French chemist Georges Urbain (1872–1938) and German chemist Carl Auer von Welsbach (1858–1929) independently of each other in 1907. There had been many false alarms because of confusion with discoveries of isotopes of the known elements. Now that elements could be unambiguously identified, it was only a matter of time before the table was completed.

Rhenium, named after the river Rhine, was discovered in platinum ores and columbite by German chemists Walter Noddack (1893–1960), Ida Tacke (1896–1979), and Otto Berg in 1925. Improved separation techniques allowed French chemist Marguérite Perey (1909–1975) in 1939 to isolate element 87 from the decay products of uranium in pitchblende where it had previously been missed by researchers. She named it francium after the country of her birth.

Synthesizing elements

The last three elements, 43, 61, and 85 were not only missing from the table but were absent from nature as well. Where nature could not provide, science found a way. US physicist Ernest O Lawrence (1901–1958) now came into the picture. He had invented the cyclotron, a device capable of accelerating charged particles to incredible speeds and energies. In 1937 he used his invention to bombard a molybdenum target with protons and found that the material had become radioactive. He enlisted the help of Italian chemists Emilio Gino Segrè (1905–1989) and Carlo Perrier (1886–1948) who separated the radioactive section and proved that it was another element. Molybdenum, atomic number 42, had been transmuted by the addition of a proton to the missing element 43. This was the first element to have been created in a laboratory and was called technetium after the Greek word meaning 'artificial'. In 1940, Segrè and the cyclotron were again involved, this time in the synthesis of element 85, which was created by bombarding element 83, bismuth, with alpha particles. In this enterprise he worked with US physicists Dale Raymond Corson and Kenneth Ross Mackenzie. They called the element astatine after the Greek word for unstable. In 1945, the final element, 61, was separated from residues recovered from a nuclear reactor by three US chemists: Jacob A Marinsky, Lawrence E Glendenin, and Charles Dubois Coryell at Clinton Laboratories, Oak Ridge, Tennessee. They named it promethium after the mythological stealer of fire from the gods, to symbolize their discovery of the element from the nuclear fire of the reactor.

The periodic table was now complete. However, when it had first been drawn up, uranium had been thought the heaviest element in nature. The possibility of being able to transmute elements beyond this point had not been considered. Elements with atomic numbers higher than uranium are called the transuranic elements and these can now be created using devices such as the cyclotron. A leading figure in this enterprise was the US nuclear chemist Glenn Seaborg (1912–1999) who was involved in the identification of a host of transuranic elements. Between 1940 and 1957 he helped in the discovery of plutonium, americium, curium, berkelium, californium, einsteinium, fermium, mendelevium, and nobelium.

Over the course of the rest of the 20th century, progressively heavier elements have been created in laboratories around the world. The last one to be created in this fashion is element 118 (ununoctium), produced by US physicists bombarding lead with krypton in 1999. It then decayed into element 116, another new element. The new elements existed only for milliseconds.

Synthesis – the rise of organic chemistry

The primary role of chemistry has always been to provide society with the compounds that it requires. In the early days of the chemical industry this usually involved the processing of natural materials from plants and minerals. However, 19th-century chemists had begun to understand the chemistry involved in making compounds containing carbon. This science is called organic chemistry, from the time that it was thought that only living, or organic, organisms could synthesize compounds of this element. The importance of this field cannot be stressed enough when we consider that every synthetic drug, dye, perfume, plastic, vitamin, rubber and a whole host of other materials, contains this element. Understandably, a great deal of effort on the part of chemists has gone into investigating carbon chemistry and the various ways in which compounds can be synthesized. The scope of this field is truly enormous and so only a brief overview of the range of developments can be addressed here.

By the beginning of the 20th century, progress in organic synthesis had largely been restricted to the production of substitutes for natural dyes and perfumes. Then several significant advances took place.

In 1897, the Sabatier–Senderens reduction reaction was developed by French chemists Paul Sabatier (1854–1941) and Jean Baptiste Senderens (1856–1936) as a means of adding hydrogen to unsaturated hydrocarbons. Hydrocarbons contain only hydrogen and carbon and are the most important class of compounds as starting materials for organic synthesis. When a hydrocarbon possesses carbon–carbon double and triple bonds it is called unsaturated and reactions with these compounds form the basis of many synthetic routes.

In 1900, French chemist Victor Grignard (1871–1935) succeeded in creating a compound where a metal was linked directly to a carbon atom (an example of an organometallic compound) by dissolving magnesium in a number of organic halide solutions. Organomagnesium compounds, now known as Grignard reagents, became some of the most versatile compounds in organic synthesis, allowing the easy preparation of a range of materials using existing products, such as the conversion of aldehydes and ketones to alcohols and the synthesis of alkyl derivatives of halogen compounds.

In 1928, German chemists Kurt Alder (1902–1958) and Otto Diels (1876–1954) discovered a fundamental reaction which allowed the synthesis of cyclic carbon compounds. As many natural vitamins and drugs contain carbon rings in their atomic structure, this reaction allowed their synthesis for the first time. The process is now called the Diels–Alder reaction.

Organocuprates – multipurpose reagents

Further advances in finding multipurpose organometallic reagents owe much to the work of US chemist Henry Gilman (1893–1986). He systematically studied the organic chemistry of a number of metals as diverse as aluminium and uranium and discovered several new types of compounds. In 1936, he was the first to study organocuprates, organic compounds in which copper is linked directly to a carbon atom. These compounds proved to be just as versatile as Grignard reagents and were to become known as Gilman reagents. They are particularly useful in catalyzing addition reactions involving carbon double and triple bonds and in substitution reactions involving organic halide and alcohol derivatives, notably in the synthesis of drugs and perfumes.

In 1954, German chemist Georg Wittig (1897–1987) developed a method of synthesizing olefins, a class of unsaturated hydrocarbons, using a reaction now called the Wittig synthesis. The process involved the reaction of an organic carbonyl, such as an aldehyde or ketone, with an organic phosphorus compound to form a compound with a carbon–carbon double bond in its atomic structure. The reaction is particularly useful in the synthesis of vitamin D and precursors of sterols such as cholesterol.

Crown ethers

In 1967, US organic chemist Charles Pedersen (1904–1990) discovered a new class of organic reagents, the crown ethers. These are planar cyclic polyethers, which are composed of molecules with twelve carbon atoms and six oxygen atoms arranged in a crown-like structure. Pedersen had been working on synthetic rubber at US company DuPont when he noticed that one of his preparations contained an unknown impurity. This turned out to be the first crown ether.

In 1969 French chemist Jean-Marie Lehn (1939–) demonstrated that the central cavity of a crown ether would accept a metal ion. He developed the procedure further by replacing oxygen atoms in the structure with nitrogen. Two crowns could then be linked together to form a three-dimensional structure which he called a 'cryptand'. This in turn led to the development of the host–guest branch of organic chemistry, the crown ether being the host and the species placed in the central cavity being the guest.

Soon afterward, US chemist Donald Cram (1919–) designed and produced a range of complex host molecules based on cryptands called 'cryptates', which selectively recognized and bound specific guest molecules and atoms. The host–guest mechanism worked because he succeeded in matching the shape of the host cavity with that of the guest species molecule. Cram demonstrated the usefulness of the technique by dissolving an inorganic salt in an organic solvent for the first time by encasing it in a cryptate. The compounds have since been used in broad applications in organic synthesis and biochemistry to catalyse chemical reactions and transport ions through biological barriers such as cell membranes.

X-ray crystallography

Important developments were also occurring in the field of structural analysis. German physicist Max von Laue (1879–1960) had shown in 1912 that passing a narrow beam of X-rays through a crystal produced a precise pattern. Later that year, English physicist Lawrence Bragg (1890–1971) showed these patterns to be caused by diffraction of the X-rays by the atoms of the crystal. This led to the conclusion that a crystal was a substance made up of an orderly arrangement of atoms repeated infinitely throughout its structure. By 1914, Bragg had been able to work out the crystal structures of a number of inorganic compounds, including salt.

This was the beginning of X-ray crystallography as an invaluable tool to the synthetic chemist. In order to manufacture a compound it is very useful to know how its molecules are made up. This technique could provide valuable insights into the molecular structure of a substance as long as the substance could be made into a crystalline form. Advancements in the preparation of organic compounds made this possible.

US biochemist James Sumner (1887–1955) crystallized the enzyme urease in 1926. He was the first to propose that enzymes were proteins, a notion rejected at first by the scientific community. However, in 1930 US chemist John Northrop (1891–1987) crystallized a number of enzymes, including pepsin, and proved all of these to be proteins, confirming Sumner's earlier work. This was followed in 1935, when US biochemist Wendell Meredith Stanley (1904–1971) isolated the tobacco-mosaic virus in crystalline form from a pulp of infected leaves.

X-ray crystallography was soon applied to the study of vitamins, antibiotics, and progressively more complex natural molecules. Two of the greatest pioneers of this field were English biophysicist Rosalind Franklin (1920–1958) and English biochemist Dorothy Hodgkin (1910–1994).

Franklin is best known for her studies into the determination of the molecular structure of DNA. In 1952, she succeeded in the extremely difficult task of obtaining an X-ray photograph from a tiny crystal of DNA. She correctly deduced that the sugar-phosphate backbone of DNA was on the outside of the molecule. Her X-ray photographs were crucial to the elucidation of the structure of DNA the following year by English physicist Francis Crick (1916–) and US biochemist James Watson (1928–).

Hodgkin was by far the most prolific user of X-ray crystallography in structural analysis of the 20th century. She developed the technique to a stage which allowed her to determine the complete molecular structure of a compound without needing to confirm her results using organic chemical techniques, a common practice before her developments in the field. Her notable triumphs were the determination of the structures of penicillin, vitamin B_{12}, and insulin, each task more difficult than the last. To

English chemist Dorothy Hodgkin at work in her laboratory in 1964. Hulton-Deutsch Collection/CORBIS

highlight the enormity of the effort involved, Hodgkin had managed to obtain the first diffraction pattern of insulin in 1935, but it took her until 1969 to complete the structural determination for the three-dimensional shape of the molecule.

The combination of an ever increasing ability to synthesize organic chemicals and an invaluable technique for the determination of molecular structures of compounds had given chemists the basic tools they required to push forward organic synthesis to limits undreamed of in the history of chemistry. A pioneer in this field was US chemist Robert Burns Woodward (1917–1979) who synthesized a number of complex natural molecules during his career, including quinine in 1944, cholesterol in 1951, and chlorophyll in 1960. He was awarded the Nobel Prize for Chemistry in 1965 for his contributions to organic synthesis. This tradition is carried on today by chemists such as US chemist Elias James Corey (1928–   ) who developed retrosynthetic analysis. Corey used this approach to synthesize more than a hundred complex natural compounds, such as terpenes, found in plant oils, and ginkgolide B, an extract from the ginkgo tree used to treat asthma. For this development he received the Nobel Prize for Chemistry in 1990.

Biochemistry – the chemistry of life

The effectiveness of the techniques of the organic chemist are ably demonstrated by the development of vitamin and drug synthesis. By 1900, it was known that foods contained substances in trace amounts that were essential for life but it was not known what these substances were. For example, it was known that fresh fruit and vegetables would cure a patient suffering from scurvy but it was not known why. The challenge was to isolate the agent responsible for the cure. In 1912, Polish-born US biochemist Casimir Funk (1884–1967) isolated an amine compound from yeast that cured the disease beriberi. He called the compound a 'vitamine' from the Latin meaning 'life amine'. A year later, US biochemists Marguerite Davis and Elmer McCollum (1879–1967) discovered another vital substance in the fats of butter and egg yolk. McCollum named this substance fat-soluble A to distinguish it from Funk's compound which he designated as water-soluble B. In 1920, British biochemist Jack Drummond proposed that the anti-scurvy substance was a third factor. In his description of the previously discovered substances, he changed Funk's designation from 'vitamine' to simply vitamin as he correctly deduced that not all the substances were amines. He called his factor vitamin C.

Progress occurred rapidly, vitamin D was isolated by McCollum in 1920 from cod-liver oil. By 1930 it had been discovered that vitamin B was in fact a group of compounds, which we call now the B vitamin complex. From this early research vitamins E and K were further identified. In 1921 the McCollum group showed that rats fed on a diet absent in vitamin D did not develop the deficiency related disease rickets when exposed to sunlight. Biochemists guessed that sunlight converted a chemical in the body of the rats into vitamin D. In 1926 British biochemists Otto Rosenheim and T A Webster and German chemist Adolf Windaus (1876–1959) independently discovered that exposing the sterol ergosterol to sunlight produced vitamin D. Ergosterol was the first example of a provitamin, a substance converted in living organisms to produce a vitamin. Vitamin C was first isolated from cabbages in 1928 by Hungarian biochemist Albert Szent-Györgi (1893–1986) although he had not known what the substance was. Around the same time US chemist Charles Glen King (1896–1988) compared crystals of a substance isolated from lemon juice with that isolated from cabbage. He found that they both protected against scurvy. In 1929, British biochemist Thomas Moore isolated the yellow coloured carotene from egg yolks. He showed that carotene was converted in living organisms to produce vitamin A. US chemists Herbert McClean Evans (1882–1971), Gladys Anderson Emerson (1903–), and Oliver Emerson succeeded in isolating vitamin E in 1936, the absence of which in the diet, they had previously shown to cause sterility in animals. US chemists Harry

Nicholls Holmes (1879–1958) and Ruth Corbet isolated pure vitamin A in crystalline form in 1937 from the oil of fish livers.

It took until 1926 before work on the chemical structures of vitamins could begin. Vitamins occur in nature only in very small amounts and it took a great deal of patient separation chemistry just to isolate enough of the pure material for analysis. After several false starts, Japanese biochemist S Ohdake came closest to the structure of vitamin B_1 when he correctly showed in 1932 that sulphur was an integral part of the vitamin. US chemist Robert Runnels Williams (1886–1965) succeeded in determining the whole structure for vitamin B_1 in 1934. The vitamin was called thiamine after the Greek words for sulphur amine. In 1933, Charles Glen King determined the structure of vitamin C, which he called ascorbic acid from the Greek meaning 'no scurvy'.

From the 1930s onwards increasing numbers of vitamins have been discovered and isolated by chemists, including vitamin K and the B group vitamins, biotin and folic acid. Synthesis followed soon afterwards. Polish-born Swiss biochemist Tadeus Reichstein (1897–1996) and English chemist Norman Haworth (1883–1950) independently synthesized ascorbic acid in 1933, Robert Runnels Williams and his group synthesized thiamine in 1937. Vitamin A had been synthesized by 1936. Soon, all vitamins could be produced synthetically and they are now routinely used as food additives and easily available as dietary supplements.

Development of antibiotics

Impressive as the discovery of vitamins was, probably the greatest contribution of synthetic chemistry to the 20th century is the part it played in the development of antibiotics.

In 1932, German chemist Gerhard Domagk (1895–1964) discovered the first antibacterial drug, a coal tar dye called Prontosil red. In 1935, his daughter was dying from streptococcal blood poisoning. Domagk used Prontosil red to kill this bacterial infection, saving his daughter's life. The active ingredient of the dye was a sulphur-containing compound called sulphanilamide. Chemists found that a range of drugs could be synthesized from sulphanilamide by substituting different chemical groups next to the sulphur atom in the molecule, producing sulphapyridine in 1937, sulphathiazole in 1939, and sulphadiazine in 1941. The 'sulfa-drugs' became the most widely used drugs until the discovery of penicillin. Each had slightly different antibacterial properties.

Once this breakthrough had been accomplished, the search was on for more potent drugs, a path which led to the discovery of antibiotics. The first true antibiotics were developed by French-US microbiologist René Jules Dubos (1901–1982) in 1939. He had been searching for antibacterial agents in soil and had found two, gramicidin and tyrocidin. But they were overshadowed by the discovery of penicillin by Scottish bacteriologist Alexander Fleming (1881–1955). In 1929, Fleming published his results on his studies of the common bread mould, *Penicillium notatum*, which contained a substance which was lethal to germs. The discovery was ignored until years later, when Australian pathologist Howard Florey (1898–1968) and German-born British biochemist Ernst Chain (1906–1979) took an interest in the paper. In 1939, they had succeeded in isolating the active agent in the bread mould, which they called penicillin. By 1941, they had managed to isolate enough of the antibacterial agent to carry out a successful clinical trial on rats. However, Florey did not have the resources in war-torn Britain to develop the full-scale purification and production of the drug, so he went to the USA, where the necessary resources were available. The first clinical trials on humans proved a complete success and by 1943 the drug was in full commercial production.

The search for antibiotics continued and streptomycin, a powerful drug in the fight against tuberculosis, was discovered in 1943 by US bacteriologist Selman Waksman (1888–1973). Aureomycin, the first of the tetracycline broad-spectrum antibiotics, was found by US botanist Benjamin Duggar (1872–1956) and colleagues in 1944. Many more discoveries followed. Antibiotics are now the most commonly used drugs in chemotherapy.

DNA – the chemical of identity

The application of chemistry to biology – biochemistry – made major contributions to science from the 1930s onwards with the analysis and then synthesis of vitamins, hormones, and other biologically active chemicals. By the mid-20th century, biochemistry had spawned its own subdivisions. One of these is molecular biology, which is concerned with the chemistry of living molecules as they exist within cells. And of all these substances, the most significant is DNA.

German cytologist Walther Flemming (1853–1905) discovered the thread-like structures in the nuclei of cells now known as chromosomes, in the 1880s. With the rediscovery of the work of Austrian monk Gregor Mendel (1822–1884) in 1900, the US geneticist Thomas Hunt Morgan (1866–1945) began studying the chromosomes of the fruit fly *Drosophila*, destined to be come the mostly widely used experimental animal in biology. In 1908 he finally made the link between chromosomes and heredity (and earned for himself the 1933 Nobel Prize for Physiology or Medicine).

Morgan showed that chromosomes consist of strings of genes. At first, scientists thought that genes were proteins. Then in 1944 that US bacteriologist Oswald Avery (1877–1955) and his colleagues demonstrated that genes are composed of deoxyribonucleic acid, (DNA). Through the genes, DNA controls all the activities of cells – metabolism, growth, division, and the formation of eggs and sperm, the progenitors of new life and the actual stuff of heredity. DNA is truly the key chemical of life.

To gain a better understanding of the role of DNA, several scientists started work to determine its structure. In the USA, chemist Linus Pauling (1901–1994) found a protein molecule that was shaped like a helix (a long spiral resembling a screw thread). In Britain, four scientists tackled the problem. Using hydrated DNA, Rosalind Franklin obtained X-ray diffraction photographs, as did Maurice Wilkins (1916–), working independently. These seemed to indicate a spiral structure. In 1952 the photographs were studied (in England) by US biochemist James Watson (1928–), who was working on the structure of DNA with English molecular biologist Francis Crick (1916–). By 1953 they had built a model of the DNA molecule, revealing its structure to be a pair of strands in the form of a double helix linked by pairs of bases, the whole thing resembling a twisted ladder. For determining the chemistry and structure of DNA, Crick, Wilkins, and Watson shared the 1962 Nobel Prize for Physiology or Medicine. Franklin had died in 1958, before her contribution could be rewarded.

DNA turned out to be a remarkable chemical. For example, during cell division it replicates itself. It does this by 'unzipping' down the centre of the helical ladder and each half then acts as a template for the creation of a new molecule. US molecular biologists Matthew Meselson (1930–) and Franklin Stahl (1929–) demonstrated this mechanism experimentally in 1958. The molecule also acts as the carrier of hereditary information. The sequence of bases in one DNA strand determines the sequence of amino acids in proteins manufactured in cells, with a combination three bases unique for each amino acid. These sequences are the genetic codes, which were first cracked by Spanish-born US biochemist Severo Ochoa (1905–1993) and, independently, US biochemist Marshall Nirenberg (1927–) who shared the 1968 Nobel Prize for Physiology or Medicine for this work.

Every cell of every organism contains DNA and the DNA of any plant or animal, including humans, is unique to that organism. Every organism has its own individual DNA profile. So in theory a sample of DNA can be used to identify the organism from which it came. One application has been in establishing the parentage (or not) of a particular individual, because his or her DNA profile reveals some characteristics of the mother and some of the father. Forensic scientists also use these DNA fingerprints to identify a body or to prove a link between a suspect and a victim in cases of rape or murder.

The petrochemical industry – the next revolution

The lead that Germany had gained on the rest of the world in the 19th century, had granted its chemical industry the position of being a virtual monopoly. German chemists maintained this position into the 20th century by continually developing novel industrial processes.

Two substances very much in demand in the early 20th century were artificial fertilizers and explosives. Both these substances contain nitrogen, the chief source of which was Chile saltpetre, a mineral mainly composed of sodium nitrate found in the northern desert of Chile. As this had been exported since 1830, supplies were becoming scarce and expensive, and so alternatives had to be found. However, nitrogen is a very stable element and does not form compounds readily. By 1900 there were several processes for converting nitrogen from the air into soluble compounds useful for chemical production, but they required too much energy to be cost effective. It was still cheaper to dig nitrates out of the ground. Then in 1908, German chemist Fritz Haber (1868–1934) developed a new process which combined atmospheric nitrogen with hydrogen to produce ammonia. From this chemical the whole range of nitrogen-based materials could be produced. He went so far as to design a pilot plant in 1909 to prove that his method was viable. Not surprisingly he called the method the Haber process. That same year, the rights to his process were bought by the German chemical firm Badische Anilin-und-Soda-Fabrik (BASF) and the task of developing the process into a full-scale industrial operation was given to German chemical engineer Carl Bosch (1874–1940). Haber had advocated the use of high pressures and an expensive osmium catalyst, both of which posed problems for Bosch. The first problem was overcome by developing a strong enough plant to withstand the high pressures. Bosch's team then found that finely divided iron, which contained proportions of oxides of potassium, calcium, and aluminium, was just as efficient as osmium as a catalyst. Another development they pioneered was passing steam over red hot coke to produce a cheap source of hydrogen. Bosch had a fully operational plant in production by 1913. The Haber process has stood the test of time and is still used to this day.

A colleague of Haber, German chemist Friedrich Bergius (1884–1949), continued improving high-pressure industrial methods. By 1912 he had developed a pilot scheme in which a combination of high temperature, high pressure, and a catalyst was used to convert coal dust and heavy oils into kerosene and petrol. Haber, Bosch, and Bergius all received Nobel prizes for their contributions to chemistry.

However, developments in the USA were about to change the direction of the chemical industry for the first time in over a century. The US chemical industry had access to a cheap source of potential chemical feedstocks in the form of petroleum, a mixture of heavy hydrocarbon oils, and natural gas. These hydrocarbon sources are much more suitable for the production of chemicals since the liquefaction phase necessary in the processing of coal is not required.

The main obstacle in the development of chemicals from petroleum sources, or petrochemicals, was that the technology did not exist to process the heavy oils and natural gas into a form that could be used. Rudimentary processes had been developed capable of breaking down or 'cracking' some of the heavy hydrocarbons in petroleum to produce petrol and the lamp oil, kerosene, but natural gas was still burnt off at oil well sites.

Thermal cracking – the production of petrol

US chemist William Burton (1865–1954) was about to change everything. He invented the Burton thermal cracker in 1912. Thermal cracking is a process which uses heat and pressure to break down heavy petroleum oils into smaller and lighter hydrocarbon fractions. Burton's process incorporated a distillation phase which converted a higher proportion of the 'cracked' petroleum into a form that could be chemically converted into petrol. He effectively doubled the amount of fuel that could be obtained from oil overnight. As a bonus, the process also produced a proportion of unsaturated hydrocarbon gases, called 'olefins' by the industry, as a by-product. In 1915, a more efficient process using thermal cracking was developed by US chemist Jesse Dubbs.

Cracking technology developed further in 1930, when French inventor Eugène Houdry (1892–1962) developed fixed-bed catalytic cracking. His process used a series of heat exchanger reactors which incorporated a bed containing a clay catalyst. This process was able to break petroleum down into a range of much lighter hydrocarbon fractions than was possible with the thermal cracking technology and soon replaced it. By 1937 commercial 'cat crackers' were in full production.

The introduction of a catalyst by Houdry was the key to the development of the most important process used in oil refining, fluidized bed catalytic cracking. This was developed in 1939 at the Massachusetts Institute of Technology headed by US chemical engineers Warren K Lewis (1882–1975) and Edwin R Gilliand. In this process petroleum is forced through a bed of catalyst at a high enough velocity to cause the particles of the bed to be separated and suspended in the liquid. This maximizes the contact with the catalyst and produces the highest yields of petroleum fractions. It became the standard process used in oil refining.

With each advance in the refining technology, a greater variety of hydrocarbon feedstocks became available for the US chemical industry to use. From the 1930s onwards, it became recognized that the olefin gases were another useful source of chemicals, especially for the growing plastics industry. This prompted the development of the commercial recovery of olefin gases from oil refineries and natural gas sources. Refrigeration and high-pressure vessels were soon routinely used to store important gases such as ethylene, propylene, and the butylenes.

The development of the technology was paralleled by the better understanding of the chemistry of petroleum. The Bureau of Standards and the US Petroleum Institute had undertaken extensive investigations into the physical and thermal properties of pure hydrocarbons. By 1931, over 190 different substances had been identified and isolated for study.

By the 1940s virtually every chemical produced by coal-based technology could be produced cheaper and in much greater quantities by the US petrochemical industry. The rest of the world had little option but to adopt the new technology. Now petrochemicals are the dominant source of raw materials for the world's chemical industry.

Plastics – the new materials

A plastic is an artificial non-metallic material that can be shaped in almost any form. The word originates from the Greek word *plastikos* that means able to be moulded. Prior to 1900, several plastic materials had been developed but they relied on being produced, at least in part, from natural materials, such as cellulose and natural rubber, and so fall short of what we now call plastics. Plastics are in the class of compounds known as polymers, which are materials built up from a series of smaller units called monomers. Other natural polymers include proteins.

The development of polymers owes a great deal to the persistence of German chemist Hermann Staudinger (1881–1965) who flew in the face of scientific opinion to push forward his theories. Staudinger was interested in the mechanism by which monomers are connected to form a polymer, a process called polymerization. He began by studying how the natural polymer, rubber, could be formed. In 1910 he developed a method of synthesizing isoprene, the monomer from which natural rubber was made. By 1920, Staudinger had learned enough to publish his book *On Polymerization*, where he discussed various mechanisms involved in polymerization. In 1922, he coined the word 'macromolecule' to describe the long chain of isoprene units which form a molecule of natural rubber. His suggestion that the molecule was composed of tens of thousands of atoms held together with ordinary chemical bonds was not well received by the scientific community. The prevailing opinion at the time was that polymers were composed of disorderly conglomerates of small molecules. However, in 1923, Staudinger's claims were verified by Swedish chemist Theodore Svedberg (1884–1971). He had developed the ultracentrifuge, a machine capable of very efficient separation of individual molecules according to their weight. Svedberg used his ultracentrifuge to separate individual molecules of proteins and plastics, which conclusively proved the existence of macromolecules. Staudinger continued his research and in 1930 he devised a relationship between the viscosity of a polymer solution and its molecular weight, Staudinger's law. This allowed the molecular weight of a polymer to be calculated without the need for complex separation techniques. For his contributions to polymer chemistry Staudinger was awarded the Nobel prize in 1956.

The first truly synthetic plastic materials were developed in New York by Belgian-born US chemist Leo Baekeland (1863–1944) in 1909. He developed the first synthetic thermoplastic Novolak and the thermosetting plastic Bakelite based on phenol-formaldehyde resins. Bakelite could be moulded into any shape, was chemically inert and nonconducting and was used widely to make electrical appliances. In 1912 German chemist Fritz Klatte patented the manufacture of vinyl chloride and proposed a method to polymerize the molecule to produce polyvinylchloride (PVC).

However, plastic development became established only when chemists gained access to olefins, a class of compounds being produced by the growing petrochemical industry in the 1930s. These unsaturated compounds could be joined together using existing synthetic techniques to form plastics. Modification of olefins using simple chemical reactions was adopted to form new monomer units such as vinyl chloride, which could be synthesized more cheaply and in greater amounts than ever before.

The Naugatuck Chemical company in Canada was the first to enter commercial production of polystyrene in 1925 using the brand name Victron. The material was expensive to produce and its yellow colour limited its applications. It was used mainly in the manufacture of false teeth. By the 1930s, chemists at the Dow Chemical Company in USA had developed a hydrogenation process for the production of styrene. This was a cheaper route and produced a white material, which was much more commercially acceptable. Full production of polystyrene under the brand name Styron took place in 1937.

Copolymerization and commercial vinyl-based plastic

Chemists at the US company Union Carbide advanced plastic synthesis in 1933 by developing a process to add two different types of monomer unit onto a polymer chain, a technique called copolymerization. They used this method to copolymerize vinyl chloride with vinyl acetate to produce Vinylite, the first commercial vinyl-based plastic.

It was known that the addition of certain chemicals in the manufacture of plastics could modify the nature of the final product. US chemist Waldo Semon (1898–1999) enhanced the flexibility of PVC in 1926 to produce the rubber-like plastic Koroseal, which was commercially available from 1932. This sort of additive is called a plasticizer and its use overcomes earlier production problems with the material. Further additive developments to improve its heat stability allowed PVC to be used in a range of applications as diverse as wire insulation and drainage piping.

The simplest of all plastics, polyethylene, was first proposed in 1898 by

German chemist Hans von Pechmann (1850–1902) but ethylene proved very difficult to polymerize. Then in 1935, English chemist Michael Perrin working at ICI in Winington, England, succeeded in polymerizing ethylene using a high-pressure vessel. The first patent was issued in 1936 under the band name Alkathene and commercial production of low-density polyethylene (LDPE) started in 1939. The process required the reaction vessel to be under an immense pressure of 30,000 lb per square inch. In 1953 German organic chemist Karl Ziegler (1898–1973) developed a chemical catalyst that permits polyethylene to be produced at atmospheric pressure. This form of the plastic, high density polyethylene (HDPE), was much cheaper to manufacture.

In 1938 US chemist Roy J Plunkett (1910–1994) working at DuPont's Jackson laboratory in New Jersey, while studying gases related to freon, discovered that a sample of tetrafluoroethylene had spontaneously polymerized upon being frozen and compressed. This was the start of polytetrafluoroethylene (PTFE) synthesis. The plastic was more widely known by its brand name Teflon. It became commercially available in 1946 and still has wide applications as a coating agent in household goods and in the aerospace industry. This research led to the development of a family of fluoropolymers, the latest one being developed in 1972.

Progress was also being made in the production of synthetic fibres from plastics. The technology to create yarn from plastic materials had been established around the turn of the century with the development of the cellulose-based rayon fibres. Soon, these techniques were being applied to plastics.

In 1938, DuPont commercially produced the synthetic fibre nylon in Seaford, Delaware, USA. This was the culmination of years of work by US chemist and polymer pioneer Wallace Hume Carothers (1896–1937). He had achieved his breakthrough in 1935 when he had polymerized adipic acid with hexamethylenediamine using a condensation reaction at low pressure to produce a perfect synthetic fibre. Nylon was a strong lightweight material and soon replaced silk in the manufacture of parachutes and ladies' stockings.

The first polyester-based synthetic fibre was developed in England in 1941 by English chemists J R Whinfield and J T Dickson but the material was not developed because of the advent of World War II and the move towards nylon production. In 1944, US chemist E F Izard working for DuPont independently developed an alternative route for polyester fibre production, only to discover after the war that the patent for the material had been filed in England by ICI. The two companies agreed to share their technology and started commercial production of polyester fibres in 1950, ICI with Terylene, DuPont with Dacron.

The acrylic fibre story began in 1939 when German chemists Otto Bayer (1902–1982). and P Kurtz patented a process for the manufacture of acrylonitrile, the first step in the production of acrylic synthetic fibres. By 1941, the German company Bayer had found a solvent for acrylic polymers that allowed them to be dissolved and then extruded into fibres. However, chemists working for DuPont in the USA independently discovered the same solvent, allowing them to develop the technology. DuPont started commercial production of acrylic fibres in Orlon, South Carolina, in 1950. Unfortunately the material could not be dyed and had an unpleasant texture which was unpopular with consumers. This particular problem was solved by rival US company Union Carbide. In 1953 they used copolymerization technology to develop Dynel, a fibre containing 40% acrylonitrile and 60% vinyl chloride. This could be dyed and had the texture of wool. From the 1950s onwards, it became common practice to mix different synthetic fibres with each other, and even with cotton, to improve the texture of the materials.

An indicator of the adaptability of the plastics industry to changing consumer trends was the development of Lycra in 1958 by DuPont. This was an example of an elastomer or elastic polymer and stretched to accommodate movement. Lycra is now one of the most common synthetic

Woman trying on nylon stockings bought from one of the first automatic vending machines, in London in 1959. Hulton-Deutsch Collection/CORBIS

fibres used in the manufacture of modern sportswear.

Plastics are one of the most common materials that are in use today. They have applications in clothing, construction, packaging, coatings, manufactured goods, glass substitutes, and insulating materials to name just a few of their uses. Research is continuing to develop plastics with useful properties. In 1977 Japanese researcher Hideki Shirakawa (1936–) and US researchers Alan McDiarmid (1929–) and Alan Heeger (1936–), while researching into a new class of electrically conductive plastics, made the discovery that the addition of iodine vastly improved the electrical properties of the polymer. By 1981, scientists at the University of Pennsylvania were able to construct the first plastic battery. In 1988, the Dutch firm CCA Biochem developed the polymer polyactide, a biodegradable plastic which can be broken down by human metabolism. It found an immediate application as a suture thread used in surgery. Developments such as these highlight that there is still a lot of research to be carried out before we find the limits to the usefulness of this class of materials.

The chemistry of death – the chemist at war

It would be wrong to highlight the development of chemistry in the 20th century without paying some attention to the darker aspect of the science, and there can be none darker than the use of the chemist during times of war.

Every discovery can often be used to do as much harm as it does good. For example, the breakthrough of the Haber process came at just the right time to be used to supply the German army with high explosives during World War I. Likewise, Bergius's development of petrol from coal helped to fuel Hitler's war machine in World War II.

In both these conflicts, increased mechanization made rubber a vital strategic material. The only place where natural rubber could be obtained was the Malaysian peninsula, but this could not be relied upon during wartime, and so synthetic alternatives were sought. The first practical synthetic rubber polymer was produced in Germany during World War I. It was made from dimethyl butadiene and was called methyl rubber. Two grades were produced, a soft grade for tyres and a hard one for battery casings used in submarines. It was not perfect and did not hold up well to stress, but was better than nothing. In 1930 both Germany and the Soviet Union developed a synthetic rubber, called Buna rubber, from butadiene using a sodium catalyst. This had superior properties to methyl rubber but was still inferior to the natural product. German chemists researched through the 1930s and eventually produced Buna-S by copolymerizing styrene with butadiene. Copolymerization proved to be the solution to their problems and by the outbreak of World War II they had developed an effective rubber substitute in the form of Buna-N, a butadiene acrylonitrile copolymer.

The USA also required a substitute to natural rubber. In 1918 US chemist J C Patrick accidentally discovered a rubbery polymer by the condensation of dichloroethane with sodium polysulphide. This was patented and sold under the name Thiokol in 1927. It was the first synthetic rubber manufactured in the USA.

In 1931, US chemists Wallace Hume Carothers and Arnold Collins of DuPont developed a synthetic rubber by polymerizing chlorobutadiene. The material was called Duprene, but the name was changed to the more familiar Neoprene in 1937. This material was superior to natural rubber in several ways as it was more resistant to organic solvents such as petrol. It became a major source of synthetic rubber used by the USA during World War II.

Chemical warfare

Arguably, the worst misuse of chemistry in the 20th century was the development of chemical weapons. Developments in industrial production during the 19th century had meant that, for the first time in history, poisonous gases could be produced in the massive quantities required to make them a viable weapon in warfare. Phosgene had been discovered in 1812, and was intensively used in synthetic dye manufacture, and liquid chlorine had been available in industrial quantities since 1880. The danger these chemicals posed caused enough concern in the international community for leading industrial nations to sign the Hague Conventions of 1899 and 1907 that strictly prohibited the use in war of asphyxiation gases. Unfortunately, this did not stop these countries from researching and producing chemical agents, just in case.

During World War I, Germany was the first to break the spirit of the agreement by using tear gas at Neuve Chapelle on October 1914 against French troops, and on January 1915 at Bolimov against the Russians. The French retaliated at Argonne in March 1915 by using a tear gas attack of their own against German troops. Escalation was inevitable. The advanced German chemical industry had given them a lead in the chemical warfare race which was demonstrated at Ypres on 22 April 1915. German troops discharged a huge chlorine gas cloud that blew over French and Canadian positions. The Allied soldiers fled, leaving a gap in their lines of 7 km/4.5 mi. Even the German high command was surprised by the success of the weapon. However, it did not follow up on the initial attack because it did not have sufficient reserves or ammunition to hold such a huge area of land. However, after the initial shock of the gas, the psychological impact of gas attacks diminished. When chlorine was deployed again two days later,

Canadian forces had improvised rudimentary protection from the gas and easily beat off the German assault. A valuable lesson was there to be learned. Chemical warfare against a prepared enemy is invariably ineffective. The wonder weapon designed to break the deadlock in World War I was not going to be gas.

Instead of abandoning chemical weapons as ineffective, both sides tried even harder to develop even more sophisticated chemical agents. The misery and suffering reached new heights with the introduction by the German army of dichlorethyl sulphide, commonly known as mustard gas due to its brown colour and distinctive smell. This chemical was a vesicant, or blistering agent, and gas masks were no protection against this development. Contact with the gas caused severe blistering of the skin and blindness if it reached the eyes. If inhaled the gas would blister the lungs, a virtual death sentence considering the medical treatment available. On the 12 and 13 July 1917 at Ypres, the weapon was used to great effect on British troops. But it was not long before both sides had this terrible weapon and the countermeasures to it. By September 1918, Allied chemists had developed mustard gas in a form that could be used as a weapon. New strategies were also tried including the use of chloropicrin, a respiratory and vomiting agent, which was able to penetrate gas masks. An affected soldier was forced to remove his gas mask in order to breathe and was therefore totally vulnerable to the lethal agent that was always put in the same shell charge.

After the war there was an immediate reaction to abolish future use of chemical agents. The treaty of Versailles in 1919 imposed a total ban on Germany to manufacture, research in, or use chemical weapons in the future. On a darker note, an article of the treaty required that the knowledge acquired by German chemists in the development of chemical weapons was to be disclosed to the Allies. In June 1925, a better step towards disarmament was taken. A protocol was signed in Geneva prohibiting the use of chemical and biological warfare. It did not ban the production, purchase, or even possession of such agents, but did prevent their use.

The fear of large-scale gas warfare increased with the approach of World War II. Between 1935 and 1936 the Italian army used mustard gas in their invasion of Abyssinia. But the expected wide-scale use of chemicals just did not happen despite the fact that German chemists had developed the next generation of weapons. In 1936 German chemist Gerhard Schrader had found the nerve agent tabun while investigating a new insecticide. He discovered the even more deadly sarin two years later. By 1942, a full-scale plant manufacturing tabun began production. But the nerve agent was never used. Advances in weapons technology, such as the development of aircraft, made the threat of retaliation against civilian populations a powerful deterrent. World War I had proved that chemical weapons were ineffective against a force that was prepared for them and so the risks involved in using these horrific weapons were just too great.

Fortunately there have been few incidents involving the use of chemical weapons since 1918. The one notable exception was the Iran–Iraq war of 1980–88. Iraqi forces used a variety of chemical weapons including phosgene and mustard gas on a large scale against largely unprotected Iranian soldiers. International condemnation against Iraq had little effect on this policy. The Iraqi army used poisonous agents against civilian targets in Iran in 1987 and in 1988 against Kurdish civilians in the town of Hallabyah in northern Iraq. Incidents such as these, show us that the dark legacy of 1914–18 is still with us.

Silent Spring – the price of success

After nearly two centuries of worldwide chemical production it was inevitable that there would be some consequences of an ever increasing dependence and demand for the products that the chemical industry provides. US chemist Thomas Midgely (1889–1944) and Swiss chemist Paul Müller (1899–1965) provide good examples of how a discovery seen as a breakthrough by one generation can become a curse to their descendants.

Midgely developed tetraethyl lead, an organometallic additive that prevents knocking in car engines, in 1921. He went on to discover freon-12 in 1930. This is an odourless, nonflammable gas that replaced the more dangerous ammonia as a refrigerant and was also used as a propellant in aerosols. At the time these were very useful developments, but unfortunately with the increased amount of motor vehicles on the roads, the lead additives in petrol now cause a serious pollution problem. They are toxic and cumulative in the human body and can result in a number of conditions including, in extreme cases, brain damage. Freon-12 is an example of a chloroflurocarbon (CFC) and all of this class of compounds are very efficient in breaking down ozone in the upper atmosphere. This accelerates ozone hole formation and increases the danger of exposure to harmful ultraviolet radiation from the Sun. By 2000 the ozone hole over Antarctica was more than three times the size of the USA.

DDT

Müller developed the insecticide dichlorodiphenyltrichloroethane, better known as DDT, in 1939. It was cheap and easy to manufacture and while nontoxic to humans was lethal to all forms of insect. It was soon used all over the world. The World Health Organization (WHO) used it as a basis to attempt to eradicate malaria by killing every mosquito, the insect which carries the disease. For a time there were dramatic drops in deaths caused

Residents of Seoul, Korea being sprayed with DDT in 1951, to prevent the spread of typhus. The toxicity of DDT was not fully appreciated until the 1970s. Bettmann/CORBIS

by malaria, and the virtual elimination of the disease in some parts of the world. Unfortunately, this overuse of DDT highlighted the flaw in this global strategy: some mosquitoes were resistant to the chemical and over time an immune mosquito population returned. In 1962, US science writer Rachel Carson (1907–1964) wrote the influential book *Silent Spring*, in which she outlined the dangers of indiscriminate use of insecticides. She pointed out that harmless and potentially useful species were being eradicated at the same time as the undesirable insects and that in many cases, insecticides can make a problem worse by removing the natural predators, but failing to eradicate the original pest. Her book fostered a growing public interest in ecology.

Oil spills and other disasters

The sheer scale of the chemical industry causes some unique problems. The supplanting of coal-based chemicals by petrochemicals has resulted in the transport of huge quantities of crude oil around the world and it was only a matter of time before a serious accident occurred. In December 1976, the Liberian tanker *Argo Merchant* ran aground off Nantucket island and broke in two spilling 180,000 barrels of crude oil into the Atlantic ocean devastating sealife for miles in every direction. In March 1989, the tanker *Exxon Valdez* spilled oil in Alaska's Prince William Sound, the oil eventually covered an area of 12,400 sq km/4,800 sq mi, and devastated the wildlife and environment of one of the world's most unspoilt regions. But by far the largest oil spill in history was the deliberate release of millions of gallons of crude oil into the Persian Gulf in 1991 by the Iraqi forces occupying Kuwait. The heavy oils, which sank to the bottom of the Gulf and destroyed the fragile ecosystem there, will persist well into the 21st century.

Many chemical companies have plants in heavily populated areas that have no protection from leakages and the consequences can be devastating. In July 1976, a pesticides plant near Seveso, Italy, released a massive cloud of dioxin gas, one of the world's most toxic poisons. It killed thousands of domestic and farm animals in the area and led to birth defects in the nearby population. In August 1978, the USA's worst chemical waste disaster occurred in the Love canal neighbourhood of Niagara Falls, New York. The houses were built on an abandoned canal which the Hooker Chemicals and Plastics Corporation had used to dump waste chemicals into between 1947 and 1953. Dioxins, pesticides, and PCBs were included in the poisonous cocktail that leaked into the basements of the houses. The leak was discovered too late for some, as a higher incident of birth defects was found to occur for the former residents. In December 1984, the world's worst chemical accident took place in Bhopal, India. The Union Carbide pesticides plant there suffered a huge leak of toxic waste which poured out into the surrounding town, killing over 6,000 people and injuring many thousands more.

The chemical industry produces large quantities of waste materials, which in many cases are discharged straight into rivers and the sea. The dilution of the effluent by water increases the difficulty in tracing the source of the pollution. However, the effects become obvious over time. A common chemical used in the manufacture of detergents, herbicides, and paints is alkylphenolethoxylate (APEO). This breaks down in contact with water to form nonylphenol, a substance which mimics a female hormone in fish. This causes male fish to partially change into females and drastically reduces fish stocks as a consequence.

The overuse of chemicals has also produced unwanted side effects. Misuse of antibiotics has led to the development of drug-resistant strains of common bacteria. The increased use of artificial fertilizers in farming has resulted in higher levels of nitrates being present in the soil than can be removed by natural processes, with excess nitrates being washed into the local water supply. The nitrate-enriched water this produces, becomes an ideal medium for the rapid growth of algae, which reduces the oxygen content of the water, killing plants and fish.

The amount of chemicals being manufactured today is greater than at any time in history. In 1983 the US Chemical Society reported that it has 6 million chemicals on record, most of which can be produced by the chemical industry. Many have been found to have had a severe environmental impact.

The future – 'a brave new world'

A primary role of the chemist of the future will be to provide alternatives to today's problem chemicals. In 1995, US chemist F Sherwood Roland (1927–), Mexican chemist Mario Molina (1943–), and Dutch chemist Paul Crutzen (1933–) were awarded the Nobel Prize for Chemistry for explaining the mechanism by which ozone reacts with pollutants such as CFCs in the upper atmosphere. This was the first chemistry award for environmental research, but it will not be the last. The rapid development of the chemistry industry has left a legacy of problems and finding their solution is the challenge chemistry faces now. Roland and his colleagues showed that understanding the problem is the first step to solving it. Their research has led to an international effort to replace CFCs with more 'ozone friendly' materials. These measures are already having an impact. Between 1988 and 1994 the amount of CFCs released into the atmosphere by the USA decreased by over 50% and in 1996 the National Oceanic and Atmospheric Administration in Washington, DC, announced the first decline in the levels of all ozone-depleting chemicals in the air. At present the majority of alternatives to CFCs still have a reduced ozone-depleting action. Future research in this field is being directed towards the development of 'ozone-safe' refrigerants and propellants at a cost that is low enough to persuade the developing countries in Africa and Asia to abandon the proven CFC technology.

New strategies for reducing existing levels of greenhouse gases have been under development since the early 1990s. Various ways to remove the major pollutant, carbon dioxide, from the atmosphere for storage in a frozen solid form exist, but these are too expensive to implement in the scale needed to have a beneficial effect. A cheaper alternative, such as reacting the gas to form a stable, solid compound, is one approach which is being studied. This route has the advantage that the resulting compound could be dumped safely in the oceans without harming the environment.

Alternatives to oil as the primary source of chemicals are now being developed such as using 'biomass' from waste vegetation. The chemical industry already has the means to process a wide range of carbon-based materials into chemicals, but at present this process is not economically viable. However, further development of the techniques involved in processing biomass may make this route competitive with petroleum at some point in the future, with the added advantage that it is free from the pollution risks inherent with petrochemicals.

The lessons of the past century have shown that the chemist of the future cannot ignore the wider implications of new chemical developments. The introduction of any new chemical must now be carefully monitored to determine its impact on the environment. Pesticides, insecticides, and fertilizers will still have to be developed to help maintain the food supply for an ever increasing population. However, better care can be shown in future developments, such as a species-specific reagent which will leave insects useful to agriculture unscathed while removing the pests. The plastics industry has already responded to the need to reduce pollution by developing the first biodegradable plastics.

The disasters and accidents associated with the chemical industry that have occurred during the last century have had at least one good effect: they have raised public awareness of the risks that exist. Now, even the most trivial chemical accident receives global news coverage. The glare of publicity has had the effect of producing new legislation that limits industrial pollution such as effluent discharge. The risk of detection of illegal waste disposal is becoming increasingly likely due to the development of techniques able to detect minute quantities of pollutants in rivers, and companies can now be heavily fined for breaching environmental legislation.

Antibiotics have proved that chemicals can be effective in treating disease. New antibiotics are continually being developed to combat the ever-increasing strains of drug-resistant bacteria which are also developing. It will be the job of the chemist to ensure that science stays ahead of nature in this race. Many potential breakthroughs in chemotherapy still await discovery, such as an anti-viral drug or an all-purpose anti-cancer agent. The chemist in the future will be expected to find the solutions to these challenges.

The 20th century taught us that we pay a price for every discovery and breakthrough that science makes. But we have more understanding than ever before of what that price is and so have the choice of whether we pay it or not. There now exists the will and the ability to decrease the effects of pollution and contamination that are our legacy from two centuries of industrial growth. We have the capability to make a difference, and if we do, then the 21st century can be 'a brave new world'.

Dr Keith B Hutton

2 Chronology

1900

French chemist François Grignard discovers organometallic compounds that become known as Grignard reagents and find important uses in organic synthesis.

Grignard discovers a series of magnesium-based organometallic compounds that help to catalyse reactions involving the combination of carbon containing groups to organic molecules. These compounds become widely used in the production of organic chemicals, notably in new approaches to alcohol synthesis.

German chemist Friedrich Thiele discovers the red hydrocarbon compound fulvene.

German physicist Friedrich Ernst Dorn discovers the radioactive noble gas radon (atomic number 86).

Dorn is studying the radioactive element radium, when he discovers that the material gives off a gas that is also radioactive, which he calls 'radium emanation'. On closer inspection, it is found that the gas is not only a new element, but also the final member of the noble gas class of elements. It is renamed 'radon' after the element radium.

US chemist Charles Skeel Palmer develops a new way of breaking down the heavy hydrocarbons ('cracking') of petroleum to obtain petrol and other products.

US chemist David Day develops the technique of absorption chromatography for separating mixtures of petroleum compounds.

1901

English chemist Frederick Gowland Hopkins discovers tryptophan, the first of the so-called essential amino acids (which have to be supplied by diet to sustain health).

French chemist Eugène-Anatole Demarçay discovers the rare-earth element europium (atomic number 63).

Demarçay isolates the oxide of the element from a compound of samarium. He names the new element 'europium' in honour of the continent of Europe. It is eventually discovered that europium is a highly efficient neutron absorber and it is used in the control systems of nuclear reactors.

German chemist Adolf Windaus discovers ergosterol, a chemical precursor of vitamin D that is converted to vitamin D by ultraviolet light in sunlight.

German chemist Emil Fischer discovers the amino acid proline.

German chemist Wilhelm Normann discovers a process for hardening liquid fats to prevent them from turning rancid.

German engineer Carl von Linde separates liquid oxygen from liquid air. It leads to the widespread use of oxygen in industry.

Japanese-born US biochemist Jokichi Takamine first isolates the heart stimulant adrenaline (epinephrine) from the suprarenal gland. It is the first pure hormone to be isolated from natural sources.

1902

Austrian chemist Richard Adolf Zsigmondy invents the ultramicroscope.

Colloidal solutions are composed of particles which are too small to be seen using a conventional microscope, but large enough to scatter light shining through the solution. The ultramicroscope projects an intense beam of light through a colloidal solution and allows the observation of the scattered light using a microscope positioned perpendicular to the projected beam. Zsigmondy discovers that he can distinguish the motion of individual particles as bright flashes against a dark background.

French chemist Auguste Verneuil develops a method of making artificial rubies and sapphires, which are important for making bearings in watches and clocks. He also discovers a way of making artificial corundum (aluminium oxide), which is used as an industrial abrasive.

French engineer and chemist Georges Claude develops a method of producing liquid air in quantity.

German chemist Friedrich Ostwald patents the industrially important process of producing nitric acid by the catalytic oxidation of ammonia.

He uses a platinum gauze catalyst to convert ammonia gas into nitrogen dioxide, which is dissolved in water to form nitric acid. His procedure becomes known as the 'Ostwald process' and becomes the primary industrial method for the manufacture of nitric acid.

German chemists Emil Fischer and Franz Hofmeister discover that proteins are polypeptides consisting of chains of amino acids.

Polish-born French physicist Marie Curie and French chemist André-Louis Debierne isolate radium as a chloride compound.

They begin the process of refining the uranium mineral pitchblende in 1898. The task takes them four years to complete, after which time they have isolated only four ounces of the chloride compound from over seven tonnes of the mineral.

The Dow Chemical Company begins the commercial manufacture of synthetic indigo in the USA.

US chemist Arthur D Little patents rayon, an artificial fibre made from cellulose.

1903

German chemist Walther Nernst prepares the first buffer solutions, used for stabilizing acidity (pH).

1904

English chemist Frederick Kipping synthesizes the first long chain silicon compound.

His compounds are structurally equivalent to saturated hydrocarbons, except that their molecules are composed of Si–Si bonds instead of C–C bonds. This class of compounds become known as 'silicones'. They are used extensively during World War II to produce non-freezing hydraulic fluids for aircraft landing gears and as general-purpose water repellents; they still find use in these applications today.

German chemist Richard Abegg announces his 'rule of eight', which recognizes the extra stability conferred on an ion or atom that has eight electrons in its outer shell.

c. 1905

English biochemist Frederick Gowland Hopkins shows that the amino acid tryptophan and other essential amino acids cannot be manufactured from other nutrients but must be supplied in the diet.

1905

German chemist Otto Hahn discovers what he believes to be a new element that he calls radiothorium. It is actually a radioisotope now called actinium-228.

US chemist Bertram Boltwood suggests that lead is the final decay product of uranium. His work will eventually lead to the uranium–lead dating method in earth sciences.

1906

British biochemist Arthur Harden discovers that enzymes catalyse the fermentation of sugar.

Russian botanist Mikhail Semyonovich Tsvet develops chromatography to separate plant pigments.

Pigments are composed of similar organic chemicals and are difficult to separate by conventional methods. Tsvet passes solutions of pigment mixtures through a column of powdered aluminium oxide and notices that each component is attracted to the powder with a different degree of strength. As the solution is washed through the tube, individual compounds separate out into different coloured fractions. He calls the technique 'chromatography' from the Greek phrase meaning 'writing in colour'. Chromatography develops into an important separation technique for complex chemical mixtures.

1907

French chemist Georges Urbain and Austrian chemist Carl Auer Freiherr von Welsbach independently discover the rare-earth element lutetium (atomic number 71).

Urbain shows that the rare earth ytterbia can be separated into two components, 'neoytterbia' and 'lutecia', the oxides of ytterbium and lutetium respectively. Welsbach carries out the same separation independently and names the two elements 'aldebaranium' and 'cassiopeium'. However, Urbain is given credit for the discovery and names the new element 'lutetium' after 'Lutetia', the ancient Roman name for the city of Paris. This is the fourteenth element to be discovered in the mineral from Ytterby, Sweden.

German chemist Emil Fischer publishes *Researches on the Chemistry of Proteins,* in which he describes the synthesis of amino acid chains in proteins.

Fischer develops a chemical procedure for combining amino acids together to form an amino acid chain eighteen units in length, which he compares with protein molecule fragments that have been broken down by digestive enzymes. These fragments are called 'peptides' after the Greek word meaning 'to digest' and his synthetic protein chain is an example of a 'synthetic polypeptide'. He discovers that his synthetic molecules have the same chemical properties as naturally produced peptides.

US chemist Bertram Boltwood discovers a radioactive element he calls ionium, now known as thorium-230. He also uses the ratio of lead and uranium in some rocks to determine their age. He estimates his samples to be 410 million to 2.2 billion years old.

1908

Belgian-born US chemist Leo Baekeland invents the plastic Bakelite: its insulating and malleable properties, combined with the fact that it does not bend when heated, ensures it has many uses.

Baekeland reacts phenol with formaldehyde to obtain a resin that is insoluble to solvents and water and is a non-conductor of electricity. The material adopts the shape of the container it is in while hot, but cools to form a hard, easily machinable solid that remains hard when reheated – the first thermosetting synthetic plastic material. He markets the material under the tradename 'Bakelite'. This discovery marks the beginning of the modern plastics industry.

French chemist Jacques Brandenberger invents cellophane.

Brandenburger develops a process for producing sheets of the cellulose-based synthetic plastic viscose, by forcing the liquid though a series of thin slots. He perfects his method and patents his process in 1911, calling the material 'cellophane'. His method produces thin transparent sheets of plastic film, which are initially used during World War I to make eyepieces for gas masks. The material is better known for its later use as a transparent wrapping material.

French physicist Jean-Baptiste Perrin proves the existence of molecules by his observation of Brownian motion of dye particles in solution.

Perrin studies Brownian motion – the apparently random movement of particles in liquids – by adding a dye of known particle size to a cylinder of water and then observing the manner in which it settles. He notices that Brownian motion is opposing the downward passage of the particles. Perrin is able to calculate the force the water molecules exert to oppose gravity and uses an equation derived by German-born US physicist

Albert Einstein to calculate the size of a water molecule. His results convince the sceptics that molecules exist.

German chemist Fritz Haber develops a process for combining hydrogen with nitrogen from air to form ammonia, the 'Haber process'.

At this time, most nitrate compounds, used in the manufacture of artificial fertilizers and explosives, are not produced synthetically because methods for the industrial production of ammonia are too expensive. Haber advocates the use of high pressures and an osmium catalyst to facilitate a reaction between hydrogen and atmospheric nitrogen that produces ammonia. The procedure is so economical that it quickly replaces all other industrial methods for ammonia production.

Japanese chemist Kikunae Ikeda isolates monosodium glutamate (MSG) from seaweed.

1909

Danish biochemist Søren Sørensen devises the pH scale for measuring acidity and alkalinity.

Sørensen originally uses pH to denote the negative logarithm of the hydrogen ion concentration in a solution. Since hydrogen ions are always present in water based solutions, the scale can be applied equally to alkaline solutions. However, uncertainties in the measurement of hydrogen concentration lead to a redefinition of the pH number to base it on standard instrument measurements. The scale is used to monitor the extent and rate of chemical reactions and has developed into an important measurement for soil assessment and pollution monitoring.

French chemist Edouard Bénédictus patents safety glass, which does not shatter into large jagged pieces when broken.

German chemist Karl Hofmann produces synthetic rubber from butadiene.

1910

French engineer and chemist Georges Claude proposes that liquid oxygen be used for smelting iron. It is not adopted until the late 1940s.

German chemist Hermann Staudinger develops a method of synthesizing isoprene, the neonomer from which natural rubber is made.

1911

New Zealand-born British physicist Ernest Rutherford proposes the concept of the nuclear atom, in which the mass of the atom is concentrated in a nucleus occupying 1/10,000 of the diameter of the atom and which has a positive charge balanced by surrounding electrons.

Rutherford projects a beam of alpha particles at a sheet of gold foil. The majority of particles pass straight through, but a small number are severely deflected. According to the existing atomic theory, this observation is the equivalent of shooting a bullet at a piece of tissue paper and having it bounce off. Rutherford proposes a new model for the structure of the atom. His model has most of the atom's mass located in a small positively charged central nucleus circled by negatively charged electrons, the majority of the atom being empty space. The alpha particles in his experiment had passed through the empty space, except those few which collided with atomic nuclei.

Russian-born US chemist Phoebus Levene discovers D-ribose, the sugar that forms the basis of RNA.

Levene isolates a five-carbon sugar from a sample of nucleic acid, the substance from which the genetic material of a cell is made up. He calls the sugar 'ribose' and, because of this discovery, every nucleic acid containing this sugar becomes known as 'ribonucleic acid' or, more commonly, by the abbreviation, RNA.

1912

Austrian chemist Friedrich Paneth and Hungarian-born Swedish chemist Georg von Hevesy develop a method of using radioactive tracers to study chemical reactions.

English biochemist Frederick Gowland Hopkins publishes the results of his experiments that prove that 'accessory substances' (vitamins) are essential for health and growth and that their absence leads to diseases such as scurvy or beriberi.

German chemist Alfred Stock is the first to synthesize boranes; his synthesized compounds are eventually used to make high-energy additives for rocket fuel.

Also known as boron hydrides, boranes are structurally equivalent to alkanes or saturated hydrocarbons, except their molecules are composed of B–B bonds instead of C–C bonds. The lighter members of this class are highly reactive, being sensitive to air and moisture, volatile and toxic. Stock develops high-vacuum apparatus and techniques in order to study boranes.

German physicist Max von Laue demonstrates that crystals are composed of regular, repeated arrays of atoms by studying the patterns in which they diffract X-rays. It is the beginning of X-ray crystallography.

Polish-born US biochemist Casimir Funk isolates vitamin B_1 (thiamine) and coins the name 'vitamines'. This proves a vital discovery in the treatment of the disease beriberi, which is caused by a deficiency of the vitamin.

Funk is searching for the 'accessory factors' in food that were proposed by Frederick Hopkins, when he isolates an amine compound from a yeast that cures beriberi. The substance Funk has discovered is part of what becomes known as the vitamin B complex. He mistakenly thinks that all essential substances contain amine groups and so renames 'accessory factors' as 'vitamines' from the Latin for 'life amine'. This name becomes shortened to 'vitamin' when it is discovered that not all of these compounds contain amine groups.

US chemist Irving Langmuir discovers that filling tungsten-filament light bulbs with inert gases greatly prolongs their life – because it eliminates the evaporation of the tungsten.

1913

Danish physicist Niels Bohr proposes that electrons orbit the atomic nucleus in fixed orbits, thus upholding New Zealand-born British physicist Ernest Rutherford's model proposed in 1911.

Bohr suggests that electrons are fixed into circular or elliptical orbits around the nucleus called shells. The theory limits the number of electrons that can occupy each shell, with the result that every element has a unique electronic configuration. The theory is used to explain and predict chemical interactions between elements. Bohr is awarded the Nobel Prize in Physics in 1922 for this work.

English chemist Frederick Soddy proposes the theory of isotopes. Soddy notices that some of the recently discovered radioactive materials have indistinguishable radioactive properties.

These had been classified as new elements because of their different atomic weights. Soddy places them in the periodic table according to their chemical properties and discovers that several of these 'elements' occupy the same space and proposes that the atomic weight of an element can vary. He calls these forms of an element 'isotopes' from the Greek words meaning 'same position'. His study of radioactive elements and isotope theory earns Soddy the Nobel Prize for Chemistry in 1921.

English physicist Henry Gwyn-Jeffreys Moseley discovers the characteristic feature of an element, the atomic number. Moseley discovers that the X-ray spectra of the elements have a deviation that changes regularly through the periodic table.

A graph of the square root of the frequency of each radiation against a quantity he calls the atomic number of the element, gives a straight line. Atomic number is later found to correspond to the number of protons in the nucleus of the atom. This quantity is the characteristic feature of all isotopes of an element and replaces atomic weight as the property used to place elements in the periodic table.

English physicists William and Lawrence Bragg further develop X-ray crystallography by establishing laws that govern the orderly arrangement of atoms in crystals display interference and diffraction patterns. They also demonstrate the wave nature of X-rays.

The Braggs show that X-ray diffraction patterns in crystalline materials are caused by interference between the atoms of a crystal and the X-rays. This leads to the conclusion that crystals are composed of regular, repeated arrays of atoms. They develop a mathematical system to relate diffraction patterns to crystal structure, which becomes known as 'Bragg's law'. This leads to the determination of the molecular structure of complex molecules such as insulin and penicillin, allowing their synthesis.

German biochemists Leonor Michaelis and Maude Menten develop a mathematical equation describing the rate of enzyme-catalysed reactions.

German chemist Friedrich Bergius publishes *The Use of High Pressure in Chemical Actions* in which he describes his process of converting coal at high pressure to produce petrol.

Bergius's process adds hydrogen to coal or heavy oils to produce petrol and kerosene. Bergius develops an industrial pilot plant for converting coal dust into petrol, but it takes a further 12 years before the first full scale plant is in operation. The hydrogenation process is used to supply Germany with petrol during World War II. Bergius shares the Nobel Prize for Chemistry in 1931 for his development of high-pressure chemical processes.

US biochemist Elmer McCollum isolates vitamin A (retinol) and is the first to use letters in the naming of vitamins.

McCollum isolates a vitamin from the fats in butter and egg yolks and names the substance fat-soluble A to distinguish it from the water soluble compound discovered by Polish-born US biochemist Casimir Funk in 1912, which he designates as water-soluble B. The use of letters becomes the standard way of naming vitamins.

US chemist William Burton patents a thermal cracking process which doubles the yield of petrol from crude oil.

Burton's process breaks up the less volatile, heavy hydrocarbon fractions of petroleum, which are then distilled and chemically converted into petrol. This doubles the yield of petrol from crude oil. Thermal cracking methods developed from the Burton process are still used to refine crude petroleum today.

US chemists Kasimir Fajans and Otto H Göhring discover the short-lived variety of the radioactive element protactinium (protactinium-234).

1914

October

Chemical agents are used for the first time as a battlefield weapon when tear gas is used by the German forces against French soldiers at Neuve Chapelle in France.

The development of the chemistry industry has made possible the production of chemical agents in sufficient quantities to make an effective weapon. This use of a chemical agent breaks the spirit of the Hague convention of 1907. French forces retaliate against the use of chemical weapons by using tear gas in March 1915 against German soldiers at Argonne, France.

1915

April

Poison gas is used in warfare for the first time by German forces at the battle of Ypres in Belgium.

The German forces release a huge cloud of chlorine gas against French and Canadian troops opening up a gap in the Allied lines across a frontage of four and a half miles. Allied forces retaliate in September 1915 at Loos, France and release a chlorine gas cloud across German lines. The use of toxic chemicals escalates from this point and becomes a common feature of World War I battle tactics. Tens of thousands of casualties on both sides result from the use of chemical agents.

1916

Dutch chemist Peter Debye demonstrates that the powdered form of a substance can be used instead of its crystal form for the X-ray study of its crystal structure.

Russian-born British (later Israeli) chemist and politician Chaim Weizmann develops a method of making acetone from grain, to be used to make the explosive cordite.

US chemist Gilbert Newton Lewis states a new valency theory, in which electrons are shared between atoms.

Lewis suggests that the complete transfer of valence shell electrons from one element to another forms ionic bonds, but a sharing of two electrons occurs to form a covalent bond. He stresses that atoms in a compound combine in order to attain the electronic configuration closest to the valence shell of a noble gas. He publishes his theory in *Valence and the Structure of Atoms and Molecules* in 1926. This theory is still used to explain the nature of chemical bonds.

1917

German physical chemist Otto Hahn and Austrian physicist Lise Meitner independently discover the radioactive element protactinium-231 (atomic number 91).

The element is present in all uranium ores and transmutes into the element actinium by radioactive decay. Polish chemist Kasimir Fajans had identified the shorter-lived isotope protactinium-234 in 1913, but Hahn and Meitner are given credit for the discovery of the element because protactinium-231 is the isotope of the element that has the longest half-life. It is initially named 'protoactinium' meaning 'before actinium', but this name is shortened to protactinium in 1949.

The German army introduces poisonous mustard gas, which causes blistering in the lungs. It is used extensively by both sides the following year.

1918

English physicist Francis Aston invents the mass spectrometer, which allows him to separate ions or isotopes of the same element. The mass spectrometer will become an important tool in stable isotope geochemistry.

The device breaks down a compound into charged atoms and fragments, which are introduced into a vacuum chamber. Powerful magnetic and electric fields are then used to separate these particles according to their mass. The machine is capable of accurately separating different isotopes of an element. In 1922, Aston publishes 'Isotopes' in which he details the isotopic composition of over 50 elements. This study of isotopes earns Aston the Nobel Prize in Chemistry for 1922.

1919

English physicist Francis Aston discovers that isotopes have atomic masses that are always whole numbers (the 'whole number rule').

English physicist J J Thomson discovers that neon exists as a number of isotopes.

Thomson uses the recently invented mass spectrograph to study the gaseous element neon. His results prove that the element is composed of at least two isotopes, Ne-20 and Ne-22. At this time, it had been thought that isotopes were restricted to radioactive elements and their decay products. Thomson's discovery proves that the theory of isotopes could also be applied to the stable elements.

1920

Belgian-born US chemist Julius Nieuwland discovers that acetylene (ethene) molecules can be polymerized to form a rubber-like substance.

US biochemists led by Elmer McCollum discover vitamin D (cholecalciferol), which is subsequently successfully used to treat rickets.

McCollum isolates a fat-soluble substance from cod liver oil that is found to cure the deficiency disease rickets and the eye condition xerophthalmia. It is the fourth vitamin to be discovered and is called vitamin D. In 1921, the same team discover that rats, which have been deprived of vitamin D in their diet, do not develop rickets if they are exposed to sunlight. This indicates that sunlight converts a substance in the skin to vitamin D.

1921

US chemist Thomas Midgley Jr invents an anti-knocking additive for petrol.

Knocking is a problem in vehicles that is caused by premature ignition of petrol vapours in the engine cylinders. Midgley discovers that the addition of the organometallic compound, tetraethyl lead, inhibits the combustion of fuel just enough to prevent an engine from knocking. He studies a range of additives and develops a mixture of tetraethyl lead, ethylene dibromide, and ethylene dichloride that not only prevents knocking, but also improves the octane rating of the fuel. It becomes commercially available in 1925 and is adopted worldwide.

1922

German chemist Hermann Staudinger discovers that rubber is a natural polymeric substance. He also coins the term 'macromolecule'.

US anatomist Herbert McLean Evans discovers vitamin E.

Evans finds that rats become sterile when limitations are placed on their diet. This condition is not relieved even when the four known vitamins, A, B, C, and D are introduced. However, if the animals are fed wheat germ, lettuce, or alfalfa, their fertility returns. He deduces that the introduced foodstuffs must contain an unknown vitamin, which he designates as vitamin E.

1923

Danish chemist Johannes Brønsted and British chemist Thomas Lowry simultaneously and independently propose the concept of acid–base pairs.

The theory of ionic dissociation states that an acid is a substance that breaks up in solution to liberate hydrogen ions. Brønsted and Lowry point out that hydrogen ions are protons, which cannot exist freely in solution and must therefore combine with other molecules immediately after dissociation. They call these groupings acid–base pairs, the acid being the substance that gives up a proton and the base being the one that accepts it. This broader concept clarifies and enhances the original theory.

Dutch chemist Peter Debye and German chemist Erich Hückel demonstrate that the disassociation of positive and negative ions of salts in solution is complete and not partial.

Dutch physicist Dirk Coster and the Hungarian-born Swedish physicist Georg von Hevesy discover the element hafnium (atomic number 72).

They identify the new element in a sample of the mineral zircon using an X-ray technique developed by Coster. They make their discovery in Copenhagen, Denmark and name the new element 'hafnium' after the Latin name for Copenhagen, 'Hafinia', in honour of this city.

Swedish Chemist Theodor Svedberg develops the ultracentrifuge, which he uses to show that polymers are composed of very large molecules comprising of hundreds of thousands of atoms.

A conventional centrifuge does not exert the force necessary to separate colloidal particles smaller than red blood corpuscles. Svedberg develops a machine that can exert the equivalent separating force of several hundreds of thousands of times the normal force of gravity. This allows the separation by mass of normal protein molecules.

Swedish chemist Theodor Svedberg (right) in a laboratory at Uppsala University in 1926.
Bettmann/CORBIS

1924

Czech chemist Jaroslav Heyrovský develops the polarograph, used for electrochemical analysis.

This device uses a mercury electrode that is configured to drop mercury through a solution into a mercury pool below. An electric current is passed through the solution and the potential is increased until the current reaches a stable plateau. The concentration of ions in the solution can be calculated from the height of the plateau. Heyrovský is able to use his device to determine the ion concentrations in solutions of unknown compositions.

1925

Austrian physicist Wolfgang Pauli discovers the exclusion principle, which accounts for the electronic structure of the atom and the chemical properties of the elements. Pauli's exclusion principle states that no two electrons can occupy the same state or configuration, which relates quantum theory to the observed properties of atoms.

German chemists Ida Tacke, Walter Noddack, and Otto Carl Berg discover the rare metallic element rhenium (atomic number 75).

They isolate an unknown substance from a platinum ore, which they recognize as being a new element. They also detect the element in molybdenite and spend the next three years isolating a single gram of the pure metal from over 600 kg of the mineral. They name their discovery 'rhenium' after 'Rhenus', the Latin name for the river Rhine.

1926

Dutch chemist Peter Debye proposes a method of obtaining temperatures a fraction of a degree above absolute zero by removing the magnetic field around a magnetized paramagnetic salt in contact with a liquid helium bath. Canadian-born US scientist William Giauque independently proposes the same idea the following year.

German chemist Hermann Staudinger proves that polymers are formed by chemical interaction between small monomer units.

Staudinger disproves the previously held belief that polymers are composed of random conglomerates of molecules held together by physical attractions. He shows that small molecules, called 'monomers', chemically combine to form long-chain straight polymer structures. In 1930, he develops a method for determining the molecular weight of a polymer molecule based on the polymer's viscosity, which becomes known as Staudinger's law.

Imperial Chemical Industries (ICI) is formed in Britain as a merger of companies, giving it a near monopoly in the British chemical industry.

US biochemist James Sumner crystallizes the enzyme urease. It is the first enzyme to be crystallized. Sumner's achievement demonstrates that enzymes are proteins.

Sumner is extracting the enzyme 'urease' from jack beans, when he notices that tiny crystals have formed in one of his sample fractions. He assumes that the crystals are a mixture of the enzyme and some other organic component. The crystals show strong enzyme activity, but every attempt at separating this activity from the crystals fails. He correctly concludes that he has crystallized a highly pure form of the enzyme.

1927

English chemist Nevil Sidgwick proposes that an atom's valence depends on the number of electrons in its outer shell and so establishes modern valence theory.

1928

German chemists Kurt Alder and Otto Diels develop a process to synthesize cyclic carbon compounds – this proves to be an important breakthrough in the development of synthetic rubber and plastics.

Alder and Diels discover a method that joins together compounds known as 'dienes', which are hydrocarbons that contain two unsaturated C=C double bonds, to form cyclic carbon compounds. The process is correctly called 'diene synthesis', but becomes commonly known as the Diels–Alder reaction. Many natural compounds containing such groups can now be artificially synthesized for the first time. It also proves to be an important breakthrough in the development of synthetic rubber and plastics.

Hungarian-born US biochemist Albert Szent-Györgyi is first to isolate crystals of vitamin C (ascorbic acid).

Szent-Györgyi isolates a crystalline substance form adrenal glands and cabbages, which he finds has the ability to lose or gain pairs of hydrogen atoms easily. He determines that the substance is composed of six carbon atoms and has the properties of a sugar, so he calls his discovery 'hexuronic acid' after the Greek for 'six' with the suffix 'uronic', which denotes a sugar. He does not realise that hexuronic acid is actually vitamin C, which is first identified in 1932.

Polyvinyl chloride (PVC) is developed, simultaneously, by the US companies Carbide and Carbon Corporation and DuPont and the German firm I G Farben.

Scottish bacteriologist Alexander Fleming discovers penicillin when he notices that the mould *Penicillium notatum*, which has invaded a culture of staphylococci, inhibits the growth of the bacteria.

Fleming observes that a culture of the bacterium *Staphylococcus aureus*, which he had left standing for several days, had failed to grow in areas where it had been accidentally contaminated with a green mould. He eventually identifies this as *Penicillium notatum*, a close relation to the common bread mould. His studies show that the mould is an effective antibacterial agent and kills many bacteria known to be harmful to humans. Although he publishes his findings the following year the significance of the discovery is not developed until World War II.

1929

French-born US chemist Eugene Houdry develops the fixed-bed catalytic process of cracking crude oil to obtain petrol, now known as the Houdry process.

Houdry's process uses a series of heat exchanger reactors, which incorporate a bed containing a clay catalyst over which the crude petroleum is passed. This is able to break down heavy hydrocarbon fractions into a much wider range of lighter fractions than is possible using the earlier thermal cracking technology. In particular, a higher proportion of unsaturated hydrocarbon gases are produced. These are important raw materials for the growing plastics industry.

Russian-born US chemist Phoebus Levene discovers 2-deoxyribose, the five-carbon sugar that forms the basis of deoxyribonucleic acid, DNA.

Levene is is studying nucleic acids that do not contain ribose, when he isolates the sugar, deoxyribose. This substance is identical to ribose with the exception that it has one less oxygen atom in its molecular structure. This proves that there are two types of nucleic acid, ribonucleic acid (RNA) and deoxyribonucleic acid (DNA). DNA is later found to be present in chromosomes.

US chemist William Giauque discovers that natural oxygen consists of three isotopes of masses 16, 17, and 18. Since oxygen is used as the standard against which relative atomic mass is determined, this means that all atomic weights are inaccurate.

The problem of choosing a new standard is not solved until 1961, when carbon-12 is selected by international agreement as the new standard weight.

1930

British chemist William Chalmers invents acrylic plastic or plexiglas (Perspex).

Dutch chemist Peter Debye investigates the structure of molecules with X-rays, providing information about the arrangement of atoms in molecules and the distances between atoms.

Russian-born Swiss biochemist Paul Karrer formulates the structure of beta-carotene, the precursor to vitamin A (retinol).

Karrer shows that vitamin is related to a family of plant pigments known as the 'carotenoids'. These substances give a distinctive orange colouring to plants, the most well-known of which is 'carotene', the colouring agent in carrots. Karrer shows that the structure of vitamin A is similar to that of carotene. He proves his theory by synthesizing beta-carotene, which can then be converted into vitamin A.

Swedish biochemist Arne Tiselius invents electrophoresis. This process, also known as 'cataphoresis', is defined as the movement of electrically charged particles through a liquid under the influence of an applied electric field.

Tiselius develops the technique to separate proteins in suspension, based on their electrical charge.

Swiss biochemist Paul Karrer formulates the structure of betacarotene, the precursor to vitamin A.

US biochemist Edward Doisy crystallizes the hormone oestriol, the first oestrogen hormone to be crystallized.

US biochemist John Northrop crystallizes pepsin and trypsin, demonstrating that they are proteins.

US chemist Thomas Midgley Jr discovers the chlorofluorocarbon (CFC) Freon–12.

Midgely discovers a process to produce the nonflammable, odourless gas di-fluoro-di-chloromethane, which is marketed under the trade name 'Freon-12'. The gas replaces ammonia and sulphur dioxide as a refrigerant in household refrigerators and air conditioning systems and is eventually used as an aerosol propellant. Unfortunately, Freon-12 is later found to be an efficient ozone depleter and is been banned in the USA and Europe to limit damage to the Earth's ozone layer.

1931

US chemist Harold C Urey and atomic physicist J Washburn discover that electrolysed water is denser than ordinary water, leading to the discovery of deuterium ('heavy hydrogen'). The discovery ushers in the modern field of stable isotope geochemistry.

Urey evaporates liquid hydrogen to concentrate any heavy isotope of hydrogen that may be present and examines the residue spectroscopically. He identifies a set of spectral lines which are slightly different than those produced by normal hydrogen. He names this new isotope 'deuterium' for the Greek word *deuteros*, which means 'second'. Deuterium has a nucleus composed of one proton and one neutron, making it double the mass of hydrogen.

US chemists of the DuPont company led by Wallace Hume Carothers develop the first commercially successful synthetic rubber, neoprene.

They base their process on work carried out by Belgian-born US chemist Julius Nieuwlands in 1920. Polymerization is a term used to describe the chemical joining of identical small molecules to form a large chain molecule, known as a polymer. The DuPont process involves the polymerization of chlorobutadiene to produce a synthetic material that is more resistant than natural rubber to organic solvents such as petrol. It is marketed as 'Duprene', but is renamed 'neoprene' in 1937.

US corporation DuPont introduces Freon, a chlorofluorocarbon (CFC), as an aerosol propellant and refrigerant; it begins to replace ammonia in refrigerators.

1932

German chemist Gerhard Domagk discovers that the red azo dye Prontosil can control streptococcal infections in mice. This is the first antibacterial sulphonamide drug ('sulfa drug').

German-born British biochemist Hans Krebs discovers the urea cycle, in which ammonia is turned into urea in mammals.

Krebs shows that amino acids degrade in mammals to form the soluble compound urea, which is then excreted from the system. This is an important step in the understanding of metabolism.

Hollywood actress sunbathing in cellophane in 1932. The cellophane was believed to allow tanning without burning. Bettmann/CORBIS

1933

Austrian-born German organic chemist Richard Kuhn isolates vitamin B_2 (riboflavin).

Kuhn announces the determination of the molecular structure of vitamin B_2 at the same time as Swiss chemist Paul Karrer, but is first to isolate the vitamin. He isolates a yellow, water-soluble compound that he calls 'riboflavin' after the Latin word meaning 'yellow'.

US biochemist Charles Glen King determines the structure of vitamin C (ascorbic acid).

King isolates the vitamin in crystalline form from cabbages and determines its molecular structure. He shows that the vitamin C molecule is made up from six carbon atoms and is similar in structure to sugars. King suggests that the isolated vitamin be called 'ascorbic acid' from the Greek phrase meaning 'no scurvy'. Szent-Györgyi, who in 1928 had isolated vitamin C but had not identified it, publishes his work within weeks of King, causing a fierce debate over who will get credit for the discovery.

English organic chemist Walter Haworth and Polish-born Swiss biochemist Tadeus Reichstein independently synthesize vitamin C (ascorbic acid).

The ability to synthesize vitamin C allows it to be mass-produced and it soon becomes a common food additive and dietary supplement.

German chemist Erich Hückel and English theoretical chemist Charles Coulson independently develop molecular orbital theory, thus founding the modern theory of how covalent bonding takes place.

1934

Australian physicist Mark Oliphant discovers tritium, the radioactive isotope of hydrogen of mass 3. He synthesizes tritium by bombarding deuterium with deuteron particles.

French physicists Frédéric and Irène Joliot-Curie bombard boron, aluminium, and magnesium with alpha particles and obtain radioactive isotopes of nitrogen, phosphorus, and aluminium – elements that are not normally radioactive. They are the first radioactive elements to be prepared artificially.

US chemist Wallace Hume Carothers develops the polyamide based synthetic fibre nylon, the first synthetic polymer fibre to be commercially produced (1938).

Carothers succeeds in polymerizing the dicarboxylic acid, adipic acid, with the diamine compound hexamethylene diamine. He uses low pressures to facilitate a condensation reaction, in which the two compounds combine alternatively to form a polymer chain. This type of polymer is called a polyamide and Carothers finds that the resultant synthetic fibre is stronger than silk. The fibre is given the trade name 'Nylon' and is commercially produced by US company DuPont in 1938.

1935

British chemist Michael Perrin and his group working for Imperial Chemical Industries (ICI) polymerize ethylene to make polyethylene, the first true plastic.

They are entering their third year of research into a process using benzaldehyde to polymerize ethylene, when they accidentally allow oxygen into the reaction vessel. This produces a small amount of the polymer, polyethylene. They correctly deduce that oxygen is essential for the polymerization reaction. The first patent for the process is issued in 1936 under the brand name 'Alkthene' and commercial production starts in 1939. This form of the plastic is low-density polyethylene (LDPE) and is used as an insulating material.

Canadian scientist William Giauque cools liquid helium to 0.0004°C/ 0.0002°F above absolute zero.

German chemist Gerhard Domagk uses the dye Prontosil red to cure a streptococcal infection in his youngest daughter; this is the first use of a sulfa drug on a human.

US biochemist Edward Calvin Kendall isolates the steroid hormone cortisone from the adrenal cortex.

US biochemist Wendell Meredith Stanley shows that viruses are not submicroscopic organisms but are proteinaceous in nature.

US biochemist William Rose discovers threonine, the last of the essential amino acids to be discovered.

US physicist Arthur Dempster discovers uranium-235, the fissionable isotope of uranium.

1936

Catalytic cracking, a chemical process in which long-chain hydrocarbon molecules are broken down into smaller ones, is introduced to produce gasoline from low grade crude oil by the US Sun Oil Company and Socony-Vacuum Company.

US physicists George Gamow and Edward Teller develop the theory of beta decay – the nuclear process of electron emission.

US chemist Robert Runnels Williams synthesizes vitamin B_1 (thiamine).

In 1933, Williams had isolated the crystalline form of vitamin B_1 and called it 'thiamine' after the Greek word for sulphur. He succeeded two years later in developing a process to synthesize the vitamin, which goes into commercial production in 1936. Williams is instrumental in the decision in the US to use thiamine as a food additive to flour, cereal grains, and corn-meal. This effectively eradicates the deficiency disease pellagra in the USA.

Vitamin E is obtained in pure form by the US biochemists Herbert Evans and Oliver and Gladys Emerson.

1937

Crystalline vitamin A and vitamin K concentrate are obtained.

Italian-born US physicist Emilio Segrè and Italian mineralogist Carlo Perrier identify technetium (atomic number 43).

US physicist Ernest Lawrence creates an unknown radioactive substance by bombarding molybdenum (atomic number 42) with the atomic nuclei of deuterium atoms, (atomic number 1) using a cyclotron. He sends the material to Perrier who discovers that the cyclotron bombardment has transmuted molybdenum into element 43. They name the element 'technetium' after the Greek word meaning 'artificial'. It is the first element to be created in a laboratory.

US company Dow Chemical Company starts to produce the plastic polystyrene. It resists attack by acids, alkalis, and many solvents, is an excellent insulator, and does not absorb water; it is used to make household appliances and toys.

British biochemists W Ewins and H Phillips synthesize sulphapyridine, the second sulfa drug.

German-born British biochemist Hans Krebs describes the citric acid cycle in cells, which converts sugars, fats, and proteins into carbon dioxide, water, and energy – the 'Krebs cycle'.

Krebs studies the mechanisms involved in the metabolism of carbohydrates. He discovers a cyclic metabolic process that begins and ends with citric acid, and shows that sugar molecules enter the system to produce carbon dioxide and hydrogen atoms. Hydrogen goes through a series of reactions with enzymes before eventually combining with oxygen to form water. This process is shown to provide the body with a source of useable energy. The process becomes known as the 'citric acid cycle' as well as the 'Krebs cycle'.

US biochemist Conrad Arnold Elvehjem finds that vitamin B_3 (niacin) prevents pellagra, a vitamin deficiency disease.

1938–40

Five independent researchers isolate and synthesize vitamin B_6 in Germany.

1938

6 April

The German physicists Lise Meitner, Otto Hahn, and Fritz Strassmann repeat Enrico Fermi's experiments and conclude that bombarding uranium atoms with neutrons splits the atom and releases huge amounts of energy by the conversion of some of the mass of the uranium atom into energy.

US chemist Roy Plunkett discovers the stable and slippery substance polytetrafluoroethylene (PTFE) (a synthetic resin), marketed by DuPont as Teflon. The most slippery substance known, it becomes commercially available in 1947–48 and is used for electrical insulation and to produce nonstick coatings; it is marketed in Britain as Fluon.

Plunkett accidentally discovers a white waxy residue while searching for a new refrigerant based on the gas tetrafluoroethylene. He discovers that the conditions used in his test cause the gas to polymerize to form polytetrafluoroethylene. This substance is resistant to solvents and heat and itis found that fats will not stick to a surface coated in the material.

1939

A team of US chemical engineers led by Warren K Lewis and Edward R Gilliand at the Massachusetts Institute of Technology develop fluidized bed catalytic cracking of petroleum. It proves to be the most efficient method of cracking ever to be developed and is still the primary method for refining petroleum in use today.

This method is similar to the Houdry process developed in 1929, with the exception that oil is introduced through a bed of catalyst instead of over it. The fluidized bed ensures that petroleum is in contact with the catalyst for the maximum amount of time.

British pathologist Howard Florey and German-born chemist Ernst Chain succeed in isolating the antibiotic penicillin from the bread mould *Penicillium notatum*, publishing their results the following year.

The process is delicate and time consuming and it is not until 1941 that they obtain enough penicillin to perform the first clinical trials on rats. It is a complete success. They do not have the resources in Britain to mass-produce the drug and so take their research to the USA. In 1943, penicillin is purified and produced in commercial quantities. Large-scale clinical trials on human patients are a complete success. Penicillin is still in common use as an antibiotic drug.

French chemist Marguérite Perey discovers the radioactive element francium (atomic number 87).

The element exists naturally as a decay product of actinium, which is itself is a decay product of uranium. Perey is studying the radioactive properties of actinium, when she detects an unusual radiation coming from her sample. She deduces that this radiation is being emitted by a decay product of actinium and eventually proves this to be element 87. She names the element 'francium' after the country of her birth.

Powder metallurgy, the manufacture of metal objects from powder rather than molten metal, is introduced.

Sulphathiazole, the third sulfa drug, is created. It is widely used in the fight against bacteria until the discovery of penicillin.

Swiss chemist Paul Müller synthesizes dichlorodiphenyltrichloroethane (DDT) and discovers its insecticidal properties.

DDT had been synthesized during the last century, however Müller discovers that is an effective agent against insects. It is cheap and easy to manufacture and soon becomes the most widely used insecticide in the world. DDT is used in a World Health Organization programme to eradicate the mosquito in an attempt to rid the world of the disease malaria. Unfortunately, the durability of the chemical leads to a progressive contamination of the environment and the food chain, with corresponding detrimental effects to human health. DDT is banned in the USA in 1972 and in Britain in 1984.

US biochemist Edward Albert Doisy determines the structure of vitamin K (phytomenadione) and synthesizes it.

The lack of vitamin K lowers the ability of blood to clot efficiently and leads to spontaneous haemorrhaging. This was discovered in 1929 by Danish biochemist Carl Peter Henrik Dam, who named the vitamin K after 'koagulation', the German for 'coagulation'. Doisy determines the molecular structure of the vitamin and then successfully develops a process to synthesize it.

US chemist Linus Pauling consolidates his theory of the chemical bond in *The Nature of the Chemical Bond and the Structure of Molecules and Crystals.* Pauling applies quantum mechanics to the study of chemical bonds.

Pauling applies experimental techniques ranging from X-ray diffraction to magnetic effects and heat measurements to provide data for the accurate calculation of angles between bonds and the interatomic distance of their association. He also proposes the concept of hybrid orbitals and explains the properties of covalent bonds.

1940

British chemist J T Dickson invents Terylene (Dacron in the USA); a polyester fibre made from terephthalic acid and ethylene glycol, it is more heat-resistant than nylon and more wear-resistant than rayon. The first yarn is made from the fibre the following year.

Canadian biochemist Martin Kamen discovers carbon-14. It becomes a vital tool in dating archaeological geological samples.

Carbon-14 is a naturally occurring radioactive isotope of carbon. It is continually being formed in the Earth's atmosphere by the bombardment of nitrogen-14 with neutrons produced by cosmic rays. It is highly stable and enters the biological carbon cycle by absorption of air by green plants. Kamen discovers the radioisotope while searching for radioactive isotopes of light elements. Carbon-14 has a radioactive half-life of over 5,000 years and decays to produce nitrogen. Its long half life is used to develop the method of age determination known as radiocarbon dating.

Italian-born US physicist Emilio Segrè and US physicists Dale Corson and K R Mackenzie synthesize astatine (atomic number 85).

Segrè, Corson, and Mackenzie bombard bismuth with alpha particles using the cyclotron at Berkeley, California, in a deliberate attempt to create element 85. They achieve their aim, but find that the new element is highly unstable and only has a half-life of a few hours. Accordingly they name their discovery astatine after the Greek word *astatos*, meaning 'unstable'.

US physicist J R Dunning leads a research team that uses a gaseous diffusion technique to isolate uranium-235 from uranium-238. Because uranium-235 readily undergoes fission into two atoms, and in doing so releases large amounts of energy, it is used for fuelling nuclear reactors.

US nuclear chemist Glenn Theodore Seaborg, and US physicists Edwin McMillan, Joseph W Kennedy, and Archer C Wahl discover plutonium (atomic number 94).

They bombard uranium with deuteron particles using the cyclotron at Berkeley, California and discover a new element in the radioactive decay products. They name the element 'plutonium' after the planet Pluto. Plutonium is the second of the transuranic elements. The element eventually becomes used in the manufacture of nuclear weapons and as a fuel in nuclear power reactors.

US physicist Edwin McMillan and US physical chemist Philip Abelson synthesize neptunium (atomic number 93), the first element found with an atomic number higher than that of uranium.

They use the cyclotron at Berkeley, California, to bombard uranium with neutrons and detect a source of radiation in the sample that has a half-life of only a few days. They discover that some of the uranium has been transmuted to create element 93. They name the element neptunium after the planet Neptune. Neptunium is placed higher than uranium in the periodic table and so becomes the first 'transuranic' element.

1941

German biochemist Fritz Lepmann discovers coenzyme A.

Lepmann explains the role of adenosine triphosphate (ATP) as the carrier of chemical energy from the oxidation of food to the energy consumption processes in the cells and shows that coenzyme A is an important catalyst in this process. He isolates coenzyme A in 1947 and determines its molecular structure in 1953.

US chemist Richard O Roblin discovers sulphadiazine; it becomes the most widely used sulfa drug.

Sulphadiazine is a synthetic compound and a member of the sulphonamide class of antibacterial drugs. It is effective against a wide spectrum of bacterial infections and has very few side effects. Sulphadiazine becomes the most widely used anitbacterial drug until the development of penicillin. The drug is still used to treat infections in patients who are allergic to penicillin-based antibiotics.

1943

Albert Hoffman, a Swiss research chemist, discovers the hallucinogenic properties of the drug LSD when he accidentally swallows some.

Chemists develop silicone rubber, the first inorganic 'rubber'. Chemically stable, silicones make good electrical insulation, waterproof coatings for fabric, and components for aircraft and cars.

1944

British biochemists Archer Martin and Richard Synge invent paper chromatography.

In 1942, they had improved the technique of partition chromatography by replacing starch with silica gel in the separating column. They continue to develop the technique and use a column of absorbent filter paper instead of silica gel. However, they introduce the liquid from the base of the column, forcing it to creep up through the separation medium. Unlike previous methods, the paper column can be dipped into a series of solvents, which greatly increases the efficiency of the separation process.

The role of deoxyribonucleic acid (DNA) in genetic inheritance is first demonstrated by US bacteriologist Oswald Avery, US biologist Colin M MacLeod, and US biologist Maclyn McCarthy.

They show that *Pneumococcus* bacteria that lack the ability to produce capsules can be given this ability by the transference of DNA obtained from capsule-producing bacteria. They correctly deduce that the genetic material for the bacteria is contained in the DNA molecule. This discovery opens the door to the elucidation of the genetic code and marks the beginning of the science of molecular genetics.

US chemists Robert Burns Woodward and William von Edders Doering synthesize quinine.

Their process builds up the complex molecule from simple compounds using a series of chemical reactions. This is an example of a totally artificial chemical synthesis, as it does not use any compounds that have been produced by a living organism.

US nuclear chemist Glenn Seaborg and his associates discover americium (atomic number 95) and curium (atomic number 96).

They bombard plutonium with neutrons and alpha particles using the cyclotron at Berkeley, California, to transmute it into the two new radioactive elements. They name them 'americium' after the USA and 'curium' in honour of the founders of nuclear chemistry, Pierre and Marie Curie.

1945

The chemical herbicides 2,4-D, 2,4,5,-T, and IPC are introduced; they herald a new era in chemical weed control as their high toxicity permits them to be used in dosages as low as one or two pounds per acre.

US chemists mix petrol and the aluminium salt of napththenic and palmitic acids to produce the jellied incendiary napalm.

1946

Carbon-13, an isotope of carbon, is discovered; it is used to cure metabolic diseases.

US biologists Max Delbrück and Alfred D Hershey discover recombinant DNA (deoxyribonucleic acid) when they observe that genetic material from different viruses can combine to create new viruses.

US physicists Edward Mills Purcell and Felix Bloch independently discover nuclear magnetic resonance, which is used to study the structure of pure metals and composites.

1947

Scottish chemist Alexander Todd synthesizes adenosine diphosphate (ADP) and adenosine triphosphate (ATP), the energy-containing molecules involved in cell metabolism.

US chemist and physicist Willard Libby develops carbon-14 dating.

Libby deduces that carbon-14 will be incorporated into plants during photosynthesis by the absorption of carbon dioxide. A living plant will continue to replenish any carbon-14 lost by radioactive decay, but this will stop when it dies. The amount of radioisotope present at that point will then diminish at a constant rate proportional to the half-life of carbon-14, which allows the determination of age by the measurement of how much remains. The age of any plant product, such as cloth, paper and wood, can be determined in this way.

US chemists Charles DuBois Coryell, Jacob Marinsky, and Lawrence E Glendenin, discover the radioactive element promethium (atomic number 61).

They discover the element among the waste deposits from a uranium fission nuclear reactor at the research site at Oak Ridge, Tennessee. In recognition that the element was recovered from a nuclear fire, they name it 'promethium' after Prometheus, the mythological Greek hero who stole fire from the Gods.

1948

Egyptian-born British chemist Dorothy Hodgkin starts to analyse the complex structure of vitamin B_{12} and makes the first X-ray photographs of it.

Scottish geneticist Charlotte Auerbach's studies begin the science of chemogenetics.

Sulphur from the Donora steel mill in Pennsylvania combines with moisture in the air to form a sulphuric acid fog that affects over 5,000 nearby residents and kills 22 people.

US chemist Karl Folkers and Scottish chemist Alexander Todd isolate vitamin B_{12}.

US physicists Richard Feynman and Julian S Schwinger, and Japanese physicist Sin-Itiro Tomonaga, independently develop quantum electrodynamics, the theory that accounts for the interactions between radiation, electrons, and positrons.

US zoologist Fairfield Osborne writes *Our Plundered Planet*, warning of the dangers posed by the insecticide DDT.

US chemists Lyman T Aldrich and Alfred Nier find argon from decay of potassium in four geologically old minerals, confirming predictions by German physicist Carl Friedrich von Weizsacker made in 1937. The basis for potassium-argon dating is established.

1949

Egyptian-born British chemist Dorothy Hodgkin determines the structure of penicillin, using a computer for the first time for this type of work.

US chemist Glenn Seaborg and his team discover berkelium (atomic number 97).

They bombard americium with helium particles using the cyclotron at Berkeley, California, to transmute it into the new radioactive element. Berkelium is highly unstable and has a half-life of less than six hours. They name it in honour of Berkeley, the home of the University of California.

1950

US chemist Glenn Seaborg and his colleagues at the University of California discover californium (atomic number 98).

They bombard curium with helium particles using the cyclotron at Berkeley, California, to transmute it into the new radioactive element. Californium has a half-life of 55 days, and is named in honour of the state and the university of California.

1951

US biochemist Robert Woodward synthesizes cortisone.

US chemists Linus Pauling and Robert Corey establish the helical or spiral structure of proteins.

Their study of the hydrogen bond shows that proteins form two types of molecular structure: either a pleated sheet or a helix (or a combination of the two). Their work helps in the later determination of the molecular structure of DNA in 1953.

US researchers Fred Joyner and Harry Coover discover the adhesive power of cyanoacrylate – 'superglue'.

c. 1952

Radioisotopes begin to be used extensively in scientific research, medicine, and industry; Britain becomes the chief exporter of isotopes.

1952

British biochemists Archer Martin and Anthony James develop gas chromatography, a technique for separating the elements of a gaseous compound.

Mixtures of gaseous compounds are passed through a liquid solvent or over an absorbent solid using an inert carrier gas, such as nitrogen. The individual components of the gas are carried at different speeds and so separate out. This technique is extremely sensitive and can be used to detect trace quantities of impurities in a sample.

English biophysicist Rosalind Franklin uses X-ray diffraction to study the structure of DNA. She suggests that its sugar-phosphate backbone is on the outside – an important clue that leads to the elucidation of the structure of DNA the following year.

US chemist Albert Ghiorso and colleagues at the University of California, Berkeley, discover the radioactive elements einsteinium (atomic number 99) and fermium (atomic number 100) in the radioactive debris collected by drone aeroplanes flown through the radioactive cloud from a hydrogen bomb explosion in the Pacific. They are named after the German-born US physicist Albert Einstein and the US physicist Enrico Fermi.

US chemist and physicist Robert S Mulliken develops a quantum mechanical theory to explain his molecular orbital theory of chemical bonds and structure.

A panel of five Nobel prizewinning chemists being interviewed by science editors at the 12th International Chemists' Conference in New York in 1951. Seated at the far side of the room from the left are: US chemist Peter Debye, Swedish chemist Arne Tiselius, Finnish chemist Artturi Virtanen, German chemist Adolf Butenandt, and English chemist Robert Robinson. Bettmann/CORBIS

Mulliken develops the theory of molecular orbitals and proposes that the electron configurations of atoms merge and conform to a molecular configuration when part of a molecule. Mulliken develops a quantum-mechanical theory to describe this behaviour.

1953

English biochemist Frederick Sanger determines the structure of the insulin molecule. The largest protein molecule to have its chemical structure determined to date, it is essential in the laboratory synthesis.

Sanger spends ten years studying the structure of the bovine insulin molecule before he correctly determines the exact order of all the amino acids in the peptide chain.

German chemist Karl Ziegler discovers a chemical catalyst that permits polyethylene plastics to be produced at atmospheric pressure. Previous methods required pressures of 30,000 lb per sq in.

Ziegler discovers that resins containing aluminium or titanium catalyst helps to catalyze the polymerization of ethylene. This new form of polyethylene is called high-density polyethylene and is stronger and more resistant to heat than the previous type. He shares the Nobel Prize for Chemistry in 1963 for this discovery.

US biochemist Vincent du Vigneaud synthesizes oxytoxin, the first protein hormone to be synthesized in a laboratory.

Vigneaud's studies of the hormone oxytocin show him that the protein is composed of only eight different amino acids, where most proteins are composed of hundreds of acid units. He then systematically builds up the peptide chain corresponding to oxytoxin using a series of chemical reactions. This shows that any protein can be synthetically manufactured.

25 April

English molecular biologist Francis Crick and US biologist James Watson announce the discovery of the double helix structure of DNA, the basic material of heredity. They also theorize that if the strands are separated then each can form the template for the synthesis of an identical DNA molecule. It is perhaps the most important discovery in biology.

They suggest that the DNA molecule consists of two chains of amino acids arranged as a double helix, with phosphate groups positioned on the outside of the molecule and purine and pyrimidine groups positioned inside. A crucial part of their analysis is based on X-ray photographs of DNA obtained by English biophysicist Rosalind Franklin.

1954

German chemist Georg Wittig develops a method of synthesizing chemical compounds containing unsaturated carbon bonds, a process that proves useful in the manufacture of vitamins A and D, prostaglandins, and sterols. Wittig discovers that a class of sulphur compounds known as ylids can be used in reactions with aldehydes and ketones to synthesize compounds containing C=C double bonds. The reaction becomes known as the Wittig synthesis.

Italian chemist Giulio Natta develops isotactic polymers; he polymerizes propylene to obtain polypropylene.

Natta develops catalysts discovered by Ziegler in 1953 to produce polymers in which all the side groupings attached to the monomer units are pointing in the same direction. He uses the term 'isotactic' to describe this type of stereo-specific polymer. He produces a family of stereo-specific catalysts capable of introducing regular structure to polymers of ethylene-

and butadiene-based plastics. This discovery highly influences future polymer development.

US chemist Robert Burns Woodward synthesizes the poison strychnine and the basis of LSD, lysergic acid.

The poisonous alkaloid strychnine is a highly complex molecule that is built up of seven interrelated rings of atoms. By synthesizing this compound, Woodward establishes himself as one of the greatest synthetic chemists in history. Later that year he synthesizes lysergic acid, the basis of the hallucinogenic drug LSD (lysergic acid diethylamide).

1955

Egyptian-born British chemist Dorothy Hodgkin determines the complex structure of vitamin B_{12} (cyanocobalamin) using X-ray photographs which she obtains from crystals of the vitamin.

The molecular structure of vitamin B_{12} is one of the most complex non-protein compounds and the task has taken Hodgkin over seven years to complete.

Russian-born Belgian physical chemist Ilya Prigogine describes the thermodynamics of irreversible processes.

Prigognine applies the second law of thermodynamics to complex systems. He suggests that the behaviour of complex systems cannot be predicted by theory if energy and matter are continually entering the system. He does, however, point out that the general behaviour of such systems can be predicted on the basis of statistical probabilities. This work is highly influential in the development of chaos theory.

US chemist Albert Ghiorso and colleagues at the University of California, Berkeley, discover the radioactive element mendelevium (atomic number 101). It is named after Russian chemist Dmitry Mendeleyev, the founder of the periodic table.

They create the element by bombarding an isotope of einsteinium with helium ions using the Berkeley 152-cm/60-in cyclotron. The radioactive element has a half-life of only 76 minutes and is the first element to be discovered one atom at a time.

1956

US biochemist and physician Arthur Kornberg discovers how DNA (deoxyribonucleic acid) molecules replicate, allowing him to synthesize DNA in a test tube.

Kornberg uses radioactively tagged nucleotides to discover that the bacteria *Escherichia coli* uses an enzyme, now known as DNA polymerase, to replicate DNA.

1958

US company DuPont develop the elastic polymer Lycra, subsequently commonly used in the manufacture of clothes, especially sportswear.

This class of synthetic fibre are known as elastiomers as they have highly elastic properties and retain their original shape even after stretching and deformation. Lycra is a polymeric material based on polyurethane and is usually covered in a protective layer of nylon.

April

US chemist Albert Ghiorso and colleagues at the University of California, Berkeley, discover nobelium (atomic number 102).

They use the heavy ion linear accelerator to bombard curium with carbon-12 ions to create the element. However, groups in Sweden and the USSR had already created a number of different isotopes of the element in 1957. Eventually the Swedish group are given permission to name the element, which they do so in honour of the Swedish chemist Alfred Nobel.

1959

Austrian-born British biochemist Max Perutz determines the structure of haemoglobin.

1960

British chemist G N Robinson discovers the antibiotic methicillin.

English biochemist John Kendrew, using X-ray diffraction techniques, elucidates the three-dimensional molecular structure of the muscle protein myoglobin.

Myoglobin is used in the body to store oxygen in muscle tissue and then release it when oxygen is needed.

US biochemist Robert Woodward and German biochemist Martin Strell independently synthesize chlorophyll.

1961

Carbon-12 becomes the internationally recognized standard by which all atomic weights are determined.

The previous standard, oxygen, had in 1929 been discovered to exist in a number of stable isotopes and it takes scientists over 30 years to agree on a replacement; this isotope of carbon is given the atomic weight of 12.011.

English molecular biologist Francis Crick and South African chemist Sydney Brenner discover that each base triplet on the DNA strand codes for a specific amino acid in a protein molecule.

French biochemists François Jacob and Jacques Monod discover messenger ribonucleic acid (mRNA), which transfers genetic information to the ribosomes, where proteins are synthesized.

US chemist Albert Ghiorso and colleagues at the University of California, Berkeley, discover the radioactive element lawrencium (atomic number 103).

They create the element by bombarding a californium target consisting of four different isotopes with two types of boron particles. This process produces a tiny amount of the new element, which they name in memory of the inventor of the cyclotron, US physicist Ernest Lawrence.

1962

British-born US chemist Herbert Charles Brown publishes *Hydroboration*, in which he outlines his discovery of organoboranes.

Brown synthesizes organoboranes by reacting sulphur-containing boron hydride compounds with unsaturated organic chemicals, such as alkenes, in an ether solvent at room temperature. These compounds are highly versatile reagents used in organic synthesis.

Canadian-born British chemist Neil Bartlett prepares the first compound of a noble gas.

Noble gases are known as the 'inert gases' at this time because no one has succeeded in producing one of their compounds. Bartlett correctly deduces that the heavier gases will be the most reactive. He immerses the very reactive compound platinum fluoride in xenon gas, which produces the compound xenon hexafluoroplatinate. Soon after, compounds of radon and krypton are also reported. The elements become known as the 'noble gases' from this time onwards.

Radioactive cobalt-60 begins to be used as a catalyst between ethylene and bromine to produce bromoethane (ethyl bromide). It is the first time ionizing radiation has been used commercially to initiate a chemical reaction.

US biochemist Robert Woodward synthesizes the antibiotic tetracycline.

US biologist and science writer Rachel Carson, in her book *Silent Spring*, draws attention to the dangers of chemical pesticides such as DDT.

Carson points out that harmless and potentially useful species are being eradicated at the same tie as the undesirable ones. She states that in many cases insecticides can make a problem worse by removing natural predators, while leaving the pest insect unaffected. Her book fosters a growing public interest in ecology.

1963

US biochemist Robert Woodward synthesizes the plant chemical colchicine.

1964

US chemist Albert Ghiorso and colleagues at the University of California, Berkeley, and researchers at the Joint Nuclear Research Institute at Dubna, USSR, independently discover rutherfordium (atomic number 104).

The element is created by bombardment of plutonium with accelerated neon nuclei. Both groups claim credit for the discovery. The US group name the element 'rutherfordium' after the New Zealand-born British physicist Ernest Rutherford, while the Soviet group name it 'kurchatovium' after the Soviet nuclear physicist Igor Kurchatov. The element is temporarily known as 'unnilquadium', the Latin for '104'.

1965

US biochemist Robert Burns Woodward synthesizes the antibiotic cephalosporin C.

1967

Korean-born US chemist Charles John Pedersen discovers crown ethers.

Pedersen accidentally discovers this new class of organic compounds while working on a new formulation for synthetic rubber at the research laboratories of the US company DuPont. He notices that an impurity in one of his preparations shows unusual properties. He isolates the impurity to discover the first crown ether. These are two-dimensional cyclic polyethers that are composed of twelve carbon atoms and six oxygen atoms arranged in a distinctive 'crown' shape with a hollow centre.

1968

US scientist Edward Feigenbaum and US geneticist Joshua Lederberg develop DENDRAL, an expert system (which duplicates human decision-making processes) for identifying chemical substances in compounds based on the results of mass-spectrographic results. Its success spurs the development of other expert systems, especially in medicine.

1969

Egyptian-born British chemist Dorothy Hodgkin determines the molecular structure of insulin.

This important compound is used to control sugar levels in the body and is an essential drug for sufferers of diabetes. Insulin is a highly complex molecule and it has taken Hodgkin 34 years to complete her structural determination of the insulin molecule. Her work allows insulin to be artificially synthesized for the first time.

French researcher Jean-Marie Lehn invents 'cryptands', synthetic crystalline materials that can capture metal ions which, once captured, act with unique chemical and electrical properties; they are used in medicine as tracers and in the production of polymers in industry.

Lehn uses the recently discovered crown ethers to design a molecule capable of accepting a metal ion. He shows that two crown ether molecules can be linked together to form a three-dimensional structure that has a large empty cavity at its centre. This cavity will accept metal ions and this discovery leads to the development of the 'host-guest' branch of organic chemistry, where the crown ether is the 'host' and a foreign species is the 'guest'.

US chemist Donald Cram discovers 'cryptates'. They make salts soluble in organic solvents such as chloroform.

Cram uses earlier work to design a range of compounds based on a 'cryptand' structure. These complex molecules are used to study the 'host–guest' mechanism. Cram successfully synthesizes a series of reagents that selectively recognize and then bind a specific type of 'guest' molecule. His study culminates in dissolving an inorganic salt into an organic solvent, a process considered to be impossible at this time.

1970

US chemist Albert Ghiorso and colleagues at the University of California, Berkeley, discover the radioactive element dubnium (unnilpentium) (atomic number 105).

They create it by bombarding californium with a beam of nitrogen nuclei using a linear accelerator. A Soviet research group at the Joint Nuclear Research Institute at Dubna, USSR, again disputes the discovery, claiming they had created the element in 1967. The US group propose the name 'hahnium' in honour of the discoverer of nuclear fission, Otto Hahn, while the Soviet group suggest 'neilsbohrium' in honour of Danish physicist Niels Bohr. The International Union of Pure and Applied Chemistry (IUPAC) recommends the name unnilpentium. Finally, in 1997, the name dubnium is adopted, again at the recommendation of IUPAC.

1971

US chemist Robert Burns Woodward and Swiss chemist Albert Eschenmoser synthesize vitamin B_{12} (cyanocobalamin).

The molecular structure of coenzyme B_{12}, also known as cyanocobalamin, is so complex that it takes the researchers over ten years to complete the synthesis of the compound. The process involves over 100 separate chemical reactions and is too complex and costly to be used in commercial production of the vitamin. However, the chemical route and the procedures used to attain it prove to be of great importance for the synthesis of many other complex compounds, such as vitamins and antibiotics.

1972

The UK and several other European countries sign an agreement prohibiting aircraft and ships from dumping toxic and plastic waste into the Atlantic; it is the first major international agreement governing pollution of the sea.

The USA restricts the use of DDT because it is discovered that it thins the eggshells of predatory birds, lowering their reproductive rates.

US microbiologist Daniel Nathans uses a restriction enzyme that splits DNA (deoxyribonucleic acid) molecules to produce a genetic map of the monkey virus (SV40), the simplest virus known to produce cancer; it is the first application of these enzymes to an understanding of the molecular basis of cancer.

1973

US biochemists Stanley Cohen and Herbert Boyer develop the technique of recombinant DNA (deoxyribonucleic acid). Strands of DNA are cut by restriction enzymes from one species and then inserted into the DNA of another; this marks the beginning of genetic engineering.

Cohen and Boyer remove DNA from the *E. coli* bacteria and then cut it into fragments. They join different fragments to the original pieces and reinsert them into the bacteria, which then reproduces normally.

1974

English biochemist Leslie Orgel shows that RNA-replicase is not necessary for RNA (ribonucleic acid) to replicate but that zinc aids this replication. It is a reaction that could have played a role in the evolution of DNA.

German chemists Manfred Eigen and Manfred Sumper demonstrate that RNA-replicase mixed with nucleotide monomers produce molecules that replicate, mutate, and evolve.

Mexican chemist Mario Molina, Dutch chemist Paul Crutzen, and US chemist Frank Sherwood Rowland warn that chlorofluorocarbons (CFCs), used in refrigerators and as aerosol propellants, may be damaging the atmosphere's ozone layer.

The ozone layer filters out much of the Sun's ultraviolet radiation and researchers suspect there is a link between chemical pollutants and the observed deterioration of this layer. The work of Molina, Crutzen, and Rowland proves this theory and they are the first to warn of the dangers of using CFCs.

1975

Atlantic salmon return to spawn in the Connecticut River, USA, after an absence of nearly 100 years; the river was restocked in 1973 after efforts were made to end pollution.

The gel-transfer hybridization technique for the detection of specific DNA sequences is developed; it is a key development in genetic engineering.

12 October

The Toxic Substances Control Act is passed by the US Congress; it requires that the production of PCBs (polychlorinated biphenyls) be phased out within three years as they have been associated with birth defects and cancer.

1976

Soviet scientists at the Joint Institute for Nuclear Research in Dubna, USSR, announce the synthesis of bohrium (atomic number 107).

They bombard a rotating cylinder coated with bismuth with charged chromium particles and detect the new element, which lasts only a few thousandths of a second. German physicist Peter Armbruster and collegues at Darmstadt, Germany, substantiate the work in 1981. They name the new element 'bohrium' after the Danish physicist Niels Bohr.

The American Panel on Atmospheric Chemistry warns that the Earth's ozone layer may be being destroyed by chloroflurocarbons (CFCs) from spray cans and refrigeration systems.

28 August
Indian-born US biochemist Har Gobind Khorana and his colleagues announce the construction of the first artificial gene to function naturally when inserted into a bacterial cell.

In 1970, Khorana had synthesized an artificial gene from individual nucleotides but had no evidence that his compound could replace a natural gene. This later work provides the evidence he needs. He has successfully developed a process to synthesize fully functional artificial genes. This is a major breakthrough in genetic engineering.

14 October
A committee to create standard regulations governing recombinant DNA research is established by the International Council of Scientific Unions.

1977

California enforces strict antipollution legislation compelling car manufacturers to install catalytic converters that reduce exhaust emissions by 90%.

Dutch scientists discover that the wastes from some incinerators are contaminated by dioxins – chemicals thought to cause cancer.

English biochemist Frederick Sanger describes the full sequence of 5,386 bases in the DNA (deoxyribonucleic acid) of virus *phi*X174 in Cambridge, England; the first sequencing of an entire genome.

This virus is a 'bacteriophage', a virus that infects bacteria, and is the first organism to have its complete genome determined.

English biochemists Frederick Sanger and Alan Coulson, and US molecular biologists Walter Gilbert and Allan Maxam, independently develop a rapid gene-sequencing technique that uses gel electrophoresis.

This technique greatly increases the speed at which nucleotides can be added together to form synthetic proteins. It is of particular importance to the rapidly growing genetics industry.

Japanese researcher Hideki Shirakawa and US researchers Alan MacDiarmid and Alan Heeger add iodine to one of the new electrically conductive plastics, vastly improving its conductive characteristics.

In 1975, Shirakawa accidentally adds a thousand times the required amount of catalyst in the preparation of a sample of the plastic, polyacetylene. This produces a new form of the plastic that is silver in appearance. He introduces his material to MacDiarmid and Heeger. This leads to the chemical doping of polyacetylene with iodine to produce the world's first conductive plastic.

December

The US Congress passes legislation banning aerosol products that emit fluorocarbons, which are thought to erode the ozone layer.

1978

The Swiss firm Pozel invents the self-heating flask; when the top is unscrewed the fluid inside is heated by a chemical reaction in a coil at the base.

August

Toxic chemicals (PCBs, dioxins, and pesticides) leak into the basements of houses in the Love Canal neighbourhood of Niagara Falls, New York. The site, an abandoned canal, was used as a chemical waste dump by the Hooker Chemicals and Plastics Corporation 1947–53. Residents are evacuated but their long-term exposure results in high rates of chromosomal damage and birth defects. It is the worst environmental disaster involving chemical waste in US history.

1979

The first International Agreement dealing with transnational air pollution is signed by European members of the United Nations (UN).

1981

Using the electricity conducting polymers developed by Hideki Shirakawa, Alan MacDiarmid, and Alan Heeger in 1977, scientists at the University of Pennsylvania, USA, construct the first plastic battery

French researchers Claude Michel and Bernard Reveau synthesize some metallic oxides that have excellent conducting properties; the materials prove invaluable in achieving superconductivity at relatively high temperatures.

1982

US firm Applied Biosystems markets an automated gene sequencer that can sequence 18,000 DNA bases a day, compared with a few hundred a year by hand in the 1970s.

1983

The American Chemical Society reports that it has recorded 6 million known chemicals.

April

US biochemist Kary Banks Mullis invents the polymerase chain reaction (PCR). This is a method of copying genes or known sections of a DNA molecule a million times without the need for a living cell. The process makes previous techniques of DNA production obsolete.

1984

German physicist Peter Armbruster and colleagues at the Heavy Ion Research Institute (GSI), Darmstadt, Germany, synthesize the element hassium (atomic number 108).

The team irradiate lead with charged iron particles to create element 108, but it only last for a few milliseconds. They name the element after *Hassius,* the Latin for the German state 'Hess'.

A leak of toxic methyl isocyanate gas from the Union Carbide pesticide plant near Bhopal, India, kills 2,600 people and injures 300,000. In February 1989 India's Supreme Court orders Union Carbide to pay $470 million in damages.

1985

English chemist Harold Kroto and US chemists Robert Curl and Richard Smalley discover fullerenes.

They use laser supersonic cluster beam apparatus built by Smalley to vaporize graphite in an inert atmosphere. Their analysis shows that they have created a previously unknown allotrope of carbon that is not diamond or graphite. The new material is mainly composed of 60 carbon atoms linked together in 12 pentagons and 20 hexagons that fit together to form a hollow sphere. They name the material 'buckminsterfullerene' after the inventor of the geodesic dome, Richard Buckminster Fuller.

1986

1 November

Water supplies along the Rhine are contaminated and millions of fish are killed when a fire at the Sandoz pharmaceutical company near Basel, Switzerland, results in 1,000 tonnes of toxic chemicals being discharged into the river.

1987

At a conference in Montreal, Canada, an international agreement, the Montreal Protocol, is reached to limit the use of ozone-depleting chlorofluorocarbons (CFCs) by 50% by the end of the century; the agreement is later condemned by environmentalists as 'too little, too late'.

1988

Dutch firm CCA Biochem develops the biodegradable polymer polyactide; capable of being broken down by human metabolism, it is ideal for use in suturing threads, bone plates, and artificial skin.

Researchers at IBM's Almaden Research Center in San José, California, using a scanning tunnelling microscope, produce the first image of the ring structure of benzene, the simplest aromatic hydrocarbon. The image confirms the structure of the molecule envisioned by Frederick Kekulé in 1865.

US chemist J Wayne Rabelais and Indian chemist Srinandan Kasi develop diamond film. It is used as an insulator and for industrial grinding.

English chemists Harold Kroto (right) and David Walton with a model of the carbon isotope buckminsterfullerene (carbon-60). Paul Seheult; Eye Ubiquitous/CORBIS

Rabelais and Kasi develop a process to coat various substrates with a layer of microcrystalline diamond – pure carbon that adheres tightly to surfaces. The layer retains the properties of normal diamond, such as hardness, but can be produced at a fraction of the cost.

1988–94

The amount of chlorofluorocarbons released into the air in the USA is reduced by 52%.

1990

Chemists at the Louis Pasteur University, Strasbourg, France, announce the creation of nucleohelicates – compounds that mimic the double helix structure of DNA (deoxyribonucleic acid).

The British company Imperial Chemical Industries (ICI) develops the first practical biodegradable plastic, Biopal.

None of the plastics used in commodity applications, such as bottles and carrier bags will quickly degrade in the environment. Increasing awareness of environmental pollution has led to the development of photodegradable polymers. Copolymerization of ethylene and carbon monoxide produces plastics that degrade when exposed to sunlight. They are similar in appearance and properties to LDPE polyethylene. However, these polymers are still not in common use due to higher production costs and instability during initial processing.

1992

Epibatidine, a chemical extracted from the skin of an Ecuadorean frog, is identified as a member of an entirely new class of alkaloid. It is an organochlorine compound and is a powerful painkiller, about 200 times as effective as morphine.

Researchers in the USA produce the first solid compound of the rare gas helium.

They mix helium and nitrogen together and apply a pressure 77,000 times greater than atmospheric pressure. The compound is thought to be held together only by Van der Waals forces. If this theory is correct, it means that the helium/nitrogen compound is the first member in a completely new class of compounds.

1993

Chemists at the University of Cambridge, England, develop light-emitting polymers (LEPs) from the semiconducting polymer poly(*p*-phenylenevinyl)

(PPV) that emit as much light as conventional LEDs and in a variety of colours.

French and Russian chemists create a superhard material by crystallizing buckminsterfullerenes at very high pressure. The material is able to scratch diamond.

The first pictures of individual atoms, obtained by the use of a scanning tunnelling microscope, are published.

1994

October

Scientists at the Heavy Ion Research Institute (GSI) based in Darmstadt, Germany, discover ununnilium (atomic number 110).

They use a linear accelerator to fuse a beam of nickel ions with a lead target. Only one atom of the new element is created. The element has not been assigned a name and is known by the temporary designation ununnilium, the Latin word for '110'.

1995

December

Scientists at the Heavy Ion Research Institute (GSI), Darmstadt, Germany, synthesize unununium (atomic number 111).

They use a linear accelerator to fuse a beam of nickel ions with a bismuth target. Only three atoms of the new element are created. The element has not been assigned a name and is known by the temporary designation unununium from the Latin word for '111'.

1996

Japanese chemists produce the first synthetic cellulose.

The US government imposes a total ban on the use and manufacture of chlorofluorocarbons (CFCs) because of their adverse effect on the ozone layer.

February

Scientists at the Heavy Ion Research Institute (GSI), Darmstadt, Germany, synthesize ununbium (atomic number 112). A single atom is created, which lasts for a third of a millisecond.

They use a linear accelerator to fuse a beam of zinc ions with a lead target. The element has not been assigned a name yet and is known by the temporary designation ununbium the Latin word for '112'.

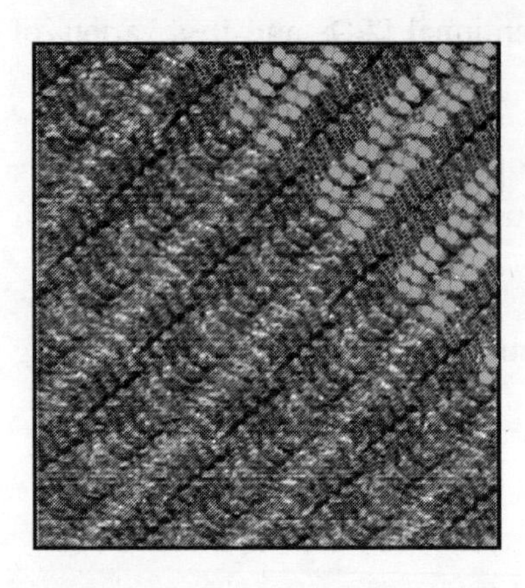

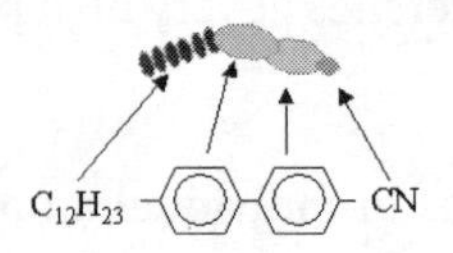

Scanning tunnelling microscopy (STM) image of a surface film of the liquid crystal (4′-dodecyl-4-cyanobiphenyl). The molecules of the liquid crystal align themselves in a regular pattern across a layer of graphite, which is determined by the size and electrical charge associated with different parts of the molecule. The crystalline layer is made up from unit cells of five pairs of molecules, indicated by the dashed lines, which are arranged in a repeatable pattern across the whole graphite surface. The high sensitivity of STM allows individual atoms to be observed. The bumps visible on the alkyl tail of the molecule ($C_{12}H_{23}$–) are hydrogen atoms and the large clump attached to this is the biphenyl group (–C_6H_6–C_6H_6–). The size of the scanned area is only 17nm^2. Courtesy of Bristol Physics SPM Group, HH Wills Physics Laboratory/AS Cherodian and M Miles.

2 June

US scientists at the National Oceanic and Atmospheric Administration in Washington, DC, announce the first decline in levels of ozone-depleting chemicals in the air.

1998

October

A study in the UK reveals that tritium pollution in the Severn Estuary in 1998 resulted in levels of tritium being hundreds of times higher than would normally be expected. Tritium is discharged into the Severn from a chemical plant in Cardiff.

October

Despite measures taken to reduce CFC emission, it is found that the ozone hole over Antarctica has increased in size to three times the area of the USA, the largest it has ever been in recorded history.

The expected size difference due to seasonal fluctuations also indicates that the hole is not regenerating as fast as it has in the past. The problem may have been exacerbated by the effects of the climatic phenomenon El Niño.

1999

March

A study published by Swiss chemists reveals that European rain contains high levels of dissolved pesticides which lead to unacceptably high pesticide levels in drinking water.

March

Russian scientists at the Joint Institute for Nuclear Research at Dubna create ununquadium (atomic number 114).

They bombard plutonium with calcium to create only a single atom of the new element. The element has not been designated a name yet and is known by the temporary assignment ununquadium, the Latin for '114'.

June

US physicists at the Lawerence Berkeley National Laboratory, USA, create ununhexium (atomic number 116) and ununoctium (atomic number 118). They use a linear accelerater to accelerate a beam of krypton ions into a lead target.

In a three-day period, they create only three atoms of ununoctium, but this element decays to produce ununhexium. The elements have not yet been designated names and are known temporarily by the Latin words for '116' and '118'

2000

27 January

The US biotech company Monsanto, at the centre of a backlash against genetically modified food, announces that it is merging with the US-Swiss pharmaceutical company Pharmacia and Upjohn to form a company worth $50 billion. The new company will be called Pharmacia.

3 Biographical Sketches

Alder, Kurt (1902–1958)

German organic chemist who with Otto Diels was awarded the Nobel Prize for Chemistry in 1950 for the discovery and development of the diene synthesis in 1928, a fundamental process that has become known as the Diels–Alder reaction. It is used in organic chemistry to synthesize cyclic (ring) compounds, including many that can be made into plastics and others – which normally occur only in small quantities in plants and other natural sources – that are the starting materials for various drugs and dyes.

Alder was born in Königshütte in Upper Silesia (now Krolewska Huta in Poland). He studied at Berlin and at Kiel, where he worked under Otto Diels. Alder was director of the Chemical Institute at the University of Cologne from 1940.

The Diels–Alder reaction involves the adding of a conjugated diene (an organic compound with two double bonds separated by a single bond) to a dienophile (a compound with only one, activated double bond). The reaction is equally general with respect to dienophiles, provided that their double bonds are activated by a nearby group such as carboxyl, carbonyl, cyano, nitro, or ester. Many of the compounds studied were prepared for the first time in Alder's laboratory.

The diene synthesis stimulated and made easier the understanding of this important group of natural products. The ease with which the reaction takes place suggests that it may be the natural biosynthetic pathway.

Baekeland, Leo Hendrik (1863–1944)

Belgian-born US chemist. He invented Bakelite, the first commercial plastic, made from formaldehyde (methanal) and phenol. He also made a photographic paper, Velox, which could be developed in artificial light.

Baekeland was born in Ghent and educated there and at Charlottenburg Technische Hochschule in Germany. He was associate professor of chemistry at Ghent University 1882–89, and later professor of chemistry and physics at Bruges. In 1889 he moved to the USA, setting up as a consultant in his own laboratory in New York. He began making Velox photographic paper in 1893, and sold the invention and the manufacturing company to Kodak for $1 million in 1899. Bakelite was patented in 1906 and in 1909 he founded the General Bakelite Corporation, later to become part of the Union Carbide and Carbon Company.

Bergius, Friedrich Karl Rudolf (1884–1949)

German chemist who invented processes for converting coal into oil and wood into sugar. He shared the Nobel Prize for Chemistry in 1931 with Carl Bosch for his part in inventing and developing high-pressure industrial methods.

Bergius was born near Breslau, Silesia (now in Poland), the son of the owner of a chemical factory. He studied chemistry at the universities of Breslau and Leipzig, and did research at Karlsruhe Technische Hochschule with German chemist Fritz Haber, who introduced him to high-pressure reactions. Bergius worked in industry 1914–45, then left Germany and eventually settled in Argentina in 1948, as a technical adviser to the government.

In 1912 Bergius worked out a pilot scheme for using high pressure, high temperature, and a catalyst to hydrogenate coal dust or heavy oil to produce paraffins (alkanes) such as petrol and kerosene. Yielding nearly 1 tonne of petrol from 4.5 tonnes of coal, the process became important to Germany during World War II as an alternative source of supply of petrol and aviation fuel. He also discovered a method of producing sugar and alcohol from simple substances made by breaking down the complex molecules in wood; he continued this work in Argentina, and found a way of making fermentable sugars and thus cattle food from wood.

Bosch, Carl (1874–1940)

German metallurgist and chemist. He developed the Haber process from a small-scale technique for the production of ammonia into an industrial high-pressure process that made use of water gas as a source of hydrogen. He shared the Nobel Prize for Chemistry in 1931 with Friedrich Bergius for his part in inventing and developing high-pressure industrial methods.

Bosch was born in Cologne and studied at Leipzig. He took a job with Badische Anilin und Sodafabrik (BASF), and by 1902 was working on methods of fixing the nitrogen present in the Earth's atmosphere. At that time, the only large sources of nitrogen compounds essential for the production of fertilizers and explosives were in the natural deposits of nitrates in Chile. Learning of the Haber process in 1908, Bosch set up a team of chemists and engineers to improve on it and reproduce it on a large scale. Heavy demand from the military during World War I caused BASF to expand. From 1925 Bosch was chair of the vast industrial conglomerate IG Farbenindustrie AG after its formation from the merger of BASF with other German industrial concerns.

Bosch's other scientific work was on the synthesis of methyl alcohol and of petrol from coal tar.

Brønsted, Johannes Nicolaus (1879–1947)

Danish physical chemist whose work in solution chemistry, particularly electrolytes, resulted in a new theory of acids and bases, the theory of proton donors and proton acceptors, published 1923.

Brønsted was born in Varde, Jutland, and studied at Copenhagen, where he became professor of physical and inorganic chemistry in 1908. In his later years he turned to politics, being elected to the Danish parliament in 1947.

Brønsted laid the foundations of the theory of the infrared spectra of polyatomic molecules by introducing the so-called valency force field in 1914. He also applied the newly developed quantum theory of specific heat capacities to gases, and published papers about the factors that determine the pH and fertility of soils.

Brønsted's new theory of acidity, published in 1923, had certain important advantages over that proposed by Swedish chemist Svante Arrhenius in 1887. Brønsted defined an acid as a proton donor and a base as a proton acceptor. The definition applies to all solvents, not just water. It also explains the different behaviour of pure acids and acids in solution. In Brønsted's scheme, every acid is related to a conjugate base, and every base to a conjugate acid.

Burton, William Merriam (1865–1954)

US inventor. Trained as a chemist, he devised the first successful 'cracking' process for yielding gasoline from crude oil, an essential step in the development of the motor industry.

He was born in Cleveland, Ohio. He worked for Standard Oil and eventually became a refinery superintendent.

Butenandt, Adolf Friedrich Johann (1903–1995)

German biochemist who was awarded the Nobel Prize for Chemistry in 1939 for his work on sex hormones. He isolated the first sex hormones (oestrone, androsterone, and progesterone), and determined their structure. He shared the Nobel prize with Swiss chemist Leopold Ružička, who synthesized androsterone.

Butenandt was born in Lehe, near Bremerhaven, and studied biology and chemistry in Marburg and Göttingen. He started to work on sex hormones while in Göttingen, and continued his research as professor of chemistry at the Danzig Institute in 1933. In 1936 he became head of the Kaiser Wilhelm Institute of Biochemistry in Berlin-Dahlem, and transferred his attentions to the study of gene action, mainly that controlling eye colour in insects.

Having determined that eye pigmentation was caused by a substance, kynurenine, formed under genetic control, Butenandt concluded in 1940 that 'genes act by providing an enzyme system that oxidizes tryptophane to kynurene.2'. This is essentially the 'one gene–one enzyme' rule usually ascribed to George Beadle. Butenandt succeeded in isolating the moulting hormone in insects – the first time an insect hormone had been isolated. He then began working on the sex attractant of the female silk moth, and thus isolated the first pheromone.

In 1960 Butenandt became president of the Max Planck Institute, and in 1972, when his presidency ended, he was elected honorary president. He remained active with the Institute for many years.

Carothers, Wallace Hume (1896–1937)

US chemist who carried out research into polymerization. He discovered that some polymers were fibre-forming, and in 1931 he produced neoprene, one of the first synthetic rubbers, and in 1935, nylon.

Carothers was born in Burlington, Iowa, and studied at Illinois and Harvard. In 1928 he became head of organic chemistry research at the DuPont research laboratory in Wilmington, Delaware. He committed suicide some years later by swallowing cyanide.

Much of Carothers's research effort was directed at producing a polymer that could be drawn out into a fibre. His first successful experiments involved polyesters, but for finer fibres with enough strength to emulate silk, he turned to polyamides. Nylon is a linear chain polymer which can be cold-drawn after extrusion through spinnerets to orientate the molecules parallel to each other so that lateral hydrogen bonding takes place.

Carothers also worked on synthetic rubbers. Neoprene, first produced commercially in 1932, is resistant to heat, light, and most solvents.

Corey, Elias James (1928–)

US organic chemist who was awarded the Nobel Prize for Chemistry in 1990 for the development of ***retrosynthetic analysis***, a method of synthesizing complex substances. A prolific worker, Corey has synthesized more than 100 substances for the first time, including terpenes (a family of hydrocarbons found in plant oils) and ginkgolide B (an extract from the ginkgo tree used to control asthma).

Retrosynthetic analysis involves the breakdown of an organic compound in stages, with each step being tested for reversibility. The starting point is a list of a compound's features and structure, such as how the carbon atoms are bonded together and whether they are linked together in chains, rings, or branches. The molecule is then simplified by unlinking the chains and breaking the bonds. From this process emerges a set of rules leading from compound to reactants and back to the compound again.

Corey was born in Methuen, Massachusetts, USA, and studied at the Massachusetts Institute of Technology, where he originally planned a career as an electrical engineer. However, an enlightening lecture by an organic chemist inspired him to change his studies to chemistry. After obtaining a doctorate, he took a job as an instructor at the University of Illinois in 1951, which led to a professorship. In 1959, he moved on to Harvard and was appointed Sheldon Emery professor in 1965.

Cornforth, John Warcup (1917–)

Australian chemist. Using radioisotopes as markers, he found out how cholesterol is manufactured in the living cell and how enzymes synthesize chemicals that are mirror images of each other (optical isomers).

He shared the Nobel Prize for Chemistry in 1975 with Swiss chemist Vladimir Prelog for his work in the stereochemistry of enzyme-catalysed reactions. He was knighted in 1977.

Cornforth was born in Sydney and educated there and at Oxford University. He settled in the UK in 1941, and worked for the British Medical Research Council 1946–62, when he became director of the Milstead Laboratory of Chemical Enzymology, Shell Research Ltd. He remained there until 1975, when he accepted a professorship at the University of Sussex.

In his researches, Cornforth studied enzymes, trying to determine specifically which group of hydrogen atoms in a biologically active compound is replaced by an enzyme to bring about a given effect. By using the element's three isotopes, he was able to identify precisely which hydrogen atom was affected by enzyme action. He was able, for example, to establish the orientation of all the hydrogen atoms in the cholesterol molecule.

Cram, Donald James (1919–)

US chemist who shared the Nobel Prize for Chemistry in 1987 with Jean-Marie Lehn and Charles J Pedersen for their work on molecules with highly selective structure-specific interactions. The work has importance in the synthesis of organic compounds, analytic chemistry, and biochemistry.

Cram designed and produced complex 'host' molecules that could selectively recognize and bind with other 'guest' molecules and ions. The molecular recognition occurs because the structure or shape of the host molecule matches that of the guest molecule. Suitable host molecules can be used to catalyse (trigger) various types of chemical reactions or to transport ions through biological barriers, such as cell membranes.

Cram was born at Chester, Vermont, USA, and educated at Rollins College, Florida, and the University of Nebraska, Canada. During World War II, he worked for Merck and Company on penicillin. In 1945, he went to Harvard University, obtaining his doctorate in 1947. After three months with the Massachusetts Institute of Technology, he joined the University of California at Los Angeles. He was appointed professor of chemistry in 1956.

Crutzen, Paul (1933–)

Dutch meteorologist who shared the Nobel Prize for Chemistry in 1995 with Mexican chemist Mario Molina and US chemist F Sherwood Rowland for their work in atmospheric chemistry, particularly concerning the formation and decomposition of ozone. They explained the chemical reactions which are destroying the ozone layer.

Crutzen, while working at Stockholm University in 1970, discovered that the nitrogen oxides NO and NO_2 speed up the breakdown of atmospheric ozone into molecular oxygen. These gases are produced in the atmosphere from nitrous oxide N_2O which is released by micro-organisms in the soil. He showed that this process is the main natural method of ozone breakdown. Crutzen also discovered that ozone-depleting chemical reactions occur on the surface of cloud particles in the stratosphere.

Crutzen was born in Amsterdam. He received his doctor's degree in meteorology from Stockholm University in 1973. He is currently at the Max Planck Institute for Chemistry in Mainz, Germany.

Curie, Marie (1867–1934)

Polish scientist, born ***Maria Sklodowska,*** who, with husband Pierre Curie, discovered in 1898 two new radioactive elements in pitchblende ores: polonium and radium. They isolated the pure elements in 1902. Both scientists refused to take out a patent on their discovery and were jointly awarded the Nobel Prize for Physics in 1903, with Henri Becquerel, for their research on radiation phenomena. Marie Curie was also awarded the Nobel Prize for Chemistry in 1911 for the discovery of radium and polonium, and the isolation and study of radium.

From 1896 the Curies worked together on radioactivity, building on the results of Wilhelm Röntgen (who had discovered X-rays) and Becquerel (who had discovered that similar rays are emitted by uranium salts). Marie Curie discovered that thorium emits radiation and found that the mineral pitchblende was even more radioactive than could be accounted for by any uranium and thorium content. In July 1898, the Curies announced the discovery of polonium, followed by the discovery of radium five months later. They eventually prepared 1 g/0.04 oz of pure radium chloride – from 8 tonnes of waste pitchblende from Austria.

They also established that beta rays (now known to consist of electrons) are negatively charged particles. In 1910 with André Debierne, who had discovered actinium in pitchblende in 1899, Marie Curie isolated pure radium metal in 1911.

Maria Sklodowska was born in Warsaw, then under Russian domination. She studied in Paris from 1891 and married in 1895. In 1906, after her husband's death, she succeeded him as professor of physics at the Sorbonne; she was the first woman to teach there. She wrote a *Treatise on Radioactivity* (1910).

At the outbreak of World War I in 1914, Curie helped to install X-ray equipment in ambulances, which she drove to the front lines. The International Red Cross made her head of its Radiological Service. Assisted by daughter Irène and Martha Klein at the Radium Institute, she held courses for medical orderlies and doctors, teaching them how to use the new technique.

By the late 1920s her health began to deteriorate: continued exposure to high-energy radiation had given her leukaemia. She and her husband had taken no precautions against radioactivity. Her notebooks, even today, are too contaminated to handle. She entered a sanatorium at Haute Savoie and died there in July 1934, a few months after her daughter and son-in-law, the Joliot-Curies, had announced the discovery of artificial radioactivity. *See illustration on page 5.*

Dam, Carl Peter Henrik (1895–1976)

Danish biochemist who was awarded a Nobel Prize for Physiology or Medicine in 1943 for his discovery of vitamin K. He shared the prize with US biochemist Edward Doisy, who received the award for determining the chemical nature of vitamin K.

In 1928 Dam began a series of experiments to see if chickens could live on a cholesterol-free diet. The birds, it turned out, were able to metabolize their own supply. Yet they continued to die from spontaneous haemorrhages. Dam concluded that their diet lacked an unknown essential ingredient to control coagulation, which he eventually found in abundance in green leaves. Dam named the new compound vitamin K.

Dam was born in Copenhagen, studied there, and spent most of his career at the university and polytechnic institute there.

Diels, Otto Paul Hermann (1876–1954)

German chemist. He and his former assistant, Kurt Alder, were awarded the Nobel Prize for Chemistry in 1950 for their research into the synthesis of organic compounds.

In 1927 Diels dehydrogenated cholesterol to produce 'Diels hydrocarbon' ($C_{18}H_{16}$), an aromatic hydrocarbon closely related to the skeletal structure of all steroids, of which cholesterol is one. In 1935 he synthesized it. This work proved to be a turning point in the understanding of the chemistry of cholesterol and other steroids.

Diels was born in Hamburg and studied at Berlin. He was director of the Chemical Institute at the Christian Albrecht University in Kiel 1916–48.

Working with his assistant Alder, Diels developed the diene synthesis, which first achieved success in 1928, when they combined cyclopentadiene with maleic anhydride (*cis*-butenedioic anhydride) to form a complex derivative of phthalic anhydride. Generally, conjugated dienes (compounds with two double bonds separated by a single bond) react with dienophiles (compounds with one double bond activated by a neighbouring substituent such as a carbonyl or carboxyl group) to form a six-membered ring.

Diels published a textbook, *Einführung in die organische Chemie/ Introduction to Organic Chemistry* (1907).

Domagk, Gerhard (1895–1964)

German pathologist who was awarded a Nobel Prize for Physiology or Medicine in 1939 for his discovery of the first antibacterial sulphonamide drug. In 1932 he found that a coal-tar dye called Prontosil red contains chemicals with powerful antibacterial properties. Sulphanilamide became the first of the sulphonamide drugs, used – before antibiotics were discovered – to treat a wide range of conditions, including pneumonia and septic wounds.

Domagk was born in Lagow, Brandenburg (now in Poland), and studied medicine at Kiel. From 1927, he directed research at the Laboratories for Experimental Pathology and Bacteriology of I G Farbenindustrie, Düsseldorf, a dyemaking company. But he also remained on the staff of Münster University, as professor from 1928.

In 1946, Domagk and his co-workers found two compounds (eventually produced under the names of Conteben and Tibione) which, although rather toxic, proved useful in treating tuberculosis caused by antibiotic-resistant bacteria.

Fischer, Hans (1881–1945)

German chemist awarded the Nobel Prize for Chemistry in 1930 for his work on haemoglobin, the oxygen-carrying, red colouring matter in blood. He determined the molecular structures of three important biological pigments: haemoglobin, chlorophyll, and bilirubin.

Fischer was born in Höchst-am-Main, near Frankfurt, and studied at Marburg and Munich. He went to Austria as professor at Innsbruck 1915–18 and Vienna 1918–21, returning to Germany as professor at the Munich Technische Hochschule. In 1945 Fischer's laboratories were destroyed in an Allied bombing raid and in a fit of despair he committed suicide.

In 1921 Fischer began investigating haemoglobin, concentrating on haem, the iron-containing non-protein part of the molecule. By 1929 he had elucidated the complete structure and synthesized haem. Chlorophyll, he found in the 1930s, has a similar structure. He then turned to the bile pigments, particularly bilirubin (the pigment responsible for the colour of the skin of patients suffering from jaundice), and by 1944 had achieved a complete synthesis of bilirubin.

Franklin, Rosalind Elsie (1920–1958)

English chemist and X-ray crystallographer who was the first to recognize the helical shape of DNA. Her work, without which the discovery of the structure of DNA would not have been possible, was built into James Watson and Francis Crick's Nobel prizewinning description of DNA.

Rosalind Franklin was born in London on 25 July 1920, into a professional family. She was educated at St Paul's School, London, and won an exhibition scholarship in 1938 to study chemistry at Cambridge University. After graduating in 1941 she stayed on to carry out postgraduate study on gas-phase chromatography with Ronald Norrish. From 1942 she studied the physical structure of coal for the British Coal Utilization Research Association, from where she moved to Paris in 1947 to research the graphitization of carbon at high temperatures at the Laboratoire Central des Services Chimique. She became a skilled X-ray crystallographer, and later applied these techniques to her study of DNA and viruses at King's College and then Birkbeck College, London.

In 1951 Franklin was appointed research associate to John Randall at King's College. Maurice Wilkins and Raymond Gosling at King's had obtained diffraction images of DNA that indicated a high degree of crystallinity and Randall gave Franklin the job of elucidating the structure of DNA.

Working with Gosling, Franklin used her chemical expertise to study the unwieldy DNA molecule and established that DNA exists in two forms – A and B – and that the sugar–phosphate backbone of DNA lies on the outside of the molecule. She also explained the basic helical structure of DNA, and produced X-ray crystallographic studies. However, she was not entirely convinced about the helical structure and was seeking further evidence in support of her theory.

Meanwhile, unknown to Randall, who had presented Franklin's data at a routine seminar, Franklin's work had found its way to her competitors Crick and Watson, at Cambridge University. They incorporated her work with that of others into their description of the double-helical structure of

DNA, which was published in 1953 in the same issue of *Nature* that Franklin's X-ray crystallographic studies of DNA were published. Franklin was not bitter about their use of her research material but began writing a corroboration of the Crick–Watson model.

Finding Wilkins difficult to work with, Franklin left King's College and joined John Bernal's laboratory at Birkbeck College to work on the tobacco mosaic virus. She died of cancer on 16 April 1958 at the age of 37, four years before she could be awarded the Nobel Prize for Physiology or Medicine with Watson, Crick, and Wilkins in 1962. The Nobel prize cannot be awarded posthumously.

Funk, Casimir (1884–1967)

Polish-born US biochemist who pioneered research into vitamins. He was the first to isolate niacin (nicotinic acid, one of the vitamins of the B complex).

Funk proposed that certain diseases are caused by dietary deficiencies. In 1912 he demonstrated that rice extracts cure beriberi in pigeons. As the extract contains an amine, he mistakenly concluded that he had discovered a class of 'vital amines', a phrase soon reduced to 'vitamins'.

Funk, born in Warsaw, studied in Berne, Switzerland, and worked at research institutes in Europe before emigrating to the USA in 1915. He returned to Warsaw in 1923 but, because of the country's uncertain political situation, went to Paris in 1927, where he founded a research institution, the Casa Biochemica. With the German invasion of France at the outbreak of World War II in 1939, Funk returned to the USA. In 1940, he became president of the Funk Foundation for Medical Research.

Funk failed to find the factor that prevents beriberi in human beings – thiamin, or vitamin B_1 – but once it had been isolated by Robert Williams, Funk determined its molecular structure in 1936 and developed a method of synthesizing it.

Funk also carried our research into animal hormones, particularly male sex hormones, and into cancer, diabetes, and ulcers. He improved methods used for drug manufacture and developed a number of new commercial products.

Gilman, Henry (1893–1986)

US organic chemist. He made a comprehensive study of methods of high-yield synthesis, quantitative and qualitative analysis, and uses of organometallic compounds, particularly Grignard reagents.

Gilman investigated the organic chemistry of 26 different metals, from aluminium, arsenic, and barium to thallium, uranium, and zinc, and discovered several new types of compounds. He was the first to study organocuprates, now known as ***Gilman reagents***, and his early work with organomagnesium compounds (Grignard reagents) would later play an important part in the preparation of polythene.

Gilman was born in Boston, Massachusetts, and studied at Harvard. He was professor at Iowa State University 1923–47.

Grignard, (François Auguste) Victor (1871–1935)

French chemist. He was awarded the Nobel Prize for Chemistry in 1912 for his discovery in 1900 of a series of organic compounds, the ***Grignard reagents***, that found applications as some of the most versatile reagents in organic synthesis. Members of the class contain a hydrocarbon radical, magnesium, and a halogen such as chlorine.

Grignard was born in Cherbourg and studied at Lyon. He became professor at Nancy in 1910. During World War I he headed a department at the Sorbonne concerned with the development of chemical warfare. From 1919 he was professor at Lyon.

Grignard reagents added to formaldehyde (methanal) produce a primary alcohol; with any other aldehyde they form secondary alcohols, and added to ketones give rise to tertiary alcohols. They will also add to a carboxylic acid to produce first a ketone and ultimately a tertiary alcohol.

His multivolume *Traité de chimie organique/Treatise on Organic Chemistry* began publication in 1935.

Haber, Fritz (1868–1934)

German chemist whose conversion of atmospheric nitrogen to ammonia opened the way for the synthetic fertilizer industry. His study of the combustion of hydrocarbons led to the commercial 'cracking' or fractional distillation of natural oil (petroleum) into its components (for example, diesel fuel, petrol, and paraffin). In electrochemistry, he was the first to demonstrate that oxidation and reduction take place at the electrodes; from this he developed a general electrochemical theory. He was awarded the Nobel Prize for Chemistry in 1918 for his work on the synthesis of ammonia from its elements.

At the outbreak of World War I in 1914, Haber was asked to devise a method of producing nitric acid for making high explosives. Later he became one of the principals in the German chemical warfare effort, devizing weapons and gas masks, which led to protests against his Nobel prize in 1918.

Haber was born in Bresslau, Silesia (now Wrocław, Poland), and educated at Berlin, Heidelberg, and the Berlin Technische Hochschule. He was

German chemist Fritz Haber at his desk at Berlin University in 1919. Hulton-Deutsch Collection/CORBIS

professor at Karlsruhe 1906–11, and then was made director of the newly established Kaiser Wilhelm Institute for Physical Chemistry in Berlin. When Adolf Hitler rose to power in 1933, Haber sought exile in Britain, where he worked at the Cavendish Laboratory, Cambridge.

After World War I, Haber set himself the task of extracting gold from sea water to help to pay off the reparations demanded from Germany by the Allies. Swedish scientist Svante Arrhenius had calculated that the sea contains 8,000 million tonnes of gold. The project got as far as the fitting-out of a ship and the commencement of the extraction process, but the yields were too low and the project was abandoned in 1928.

Hevesy, Georg Karl von (1885–1966)

Hungarian-born Swedish chemist, discoverer of the element hafnium. He was awarded the Nobel Prize for Chemistry in 1943 for his use of a radioactive isotope to follow the steps of a biological process.

Hevesy was born in Budapest and educated in Germany, Switzerland, and the UK, studying at Manchester under nuclear physics pioneer Ernest Rutherford in 1911. He worked in Copenhagen at the Institute of Physics under Niels Bohr 1920–26 and 1934–43. During the German occupation of

Denmark in World War II, Hevesy escaped to Sweden and became professor at Stockholm.

Hevesy first used radioactive tracers to study the chemistry of lead and bismuth salts. During the early 1930s, he began experiments with this technique on biological specimens, noting the take-up of radioactive lead by plants, and going on to use an isotope to trace the movement of phosphorus in the tissues of the human body. He used heavy water to study the mechanism of water exchange between goldfish and their surroundings and also within the human body. Using radioactive calcium to label families of mice, he showed that, of calcium atoms present at birth, about 1 in 300 are passed on to the next generation.

Hodgkin, Dorothy Mary Crowfoot (1910–1994)

Egyptian-born British chemist who analysed the structure of penicillin, insulin, and vitamin B_{12}. Hodgkin was the first to use a computer to analyse the molecular structure of complex chemicals, and this enabled her to produce three-dimensional models. She was awarded the Nobel Prize for Chemistry in 1964 for her work in the crystallographic determination of the structures of biochemical compounds, notably penicillin and cyanocobalamin (vitamin B_{12}).

Hodgkin studied the structures of calciferol (vitamin D_2), lumisterol, and cholesterol iodide, the first complex organic molecule to be determined completely by the pioneering technique of X-ray crystallography, a physical analysis technique devised by Lawrence Bragg (1890–1971), and at the time used only to confirm formulas predicted by organic chemical techniques. She also used this technique to determine the structure of penicillin, insulin, and vitamin B_{12}.

Hodgkin was born in Cairo and educated at Somerville College, Oxford. At Cambridge (1932–34) she worked on the development of X-ray crystallography and, returning to Oxford in 1934, began working on penicillin. After Howard Florey and Ernst Chain isolated penicillin from mould in 1939, chemists in Britain and the USA raced to determine its structure. Hodgkin's assertion that the core of penicillin consisted of a ring of three carbons and a nitrogen thought too unstable to exist, brought from Australian chemist John Cornforth the derisive comment, 'If that's the formula of penicillin, I'll give up chemistry and grow mushrooms.' But Hodgkin was right, and she went on to determine the structure of the antibiotic cephalosporin C.

In 1948 Hodgkin began her work on vitamin B_{12}, a substance that proved to be far more complex than penicillin: the first X-ray diffraction pictures showed over 1,000 atoms, compared to penicillin's 39. It took Hodgkin and her co-workers eight years to solve.

The structure of insulin was to take much longer still; Hodgkin first saw the diffraction pattern made by insulin in 1935, but it was to take 34 years for her to determine its structure.

In 1964 Hodgkin became the second woman to have ever received the Order of Merit (the first was Florence Nightingale) and – a committed socialist all her life – in 1987 she was awarded the Lenin Peace Prize. She was chancellor of Bristol University 1970–88 and helped found a Hodgkin scholarship for students from the developing world.

Houdry, Eugene Jules (1892–1962)

French-born US chemical engineer and inventor. By 1927 he had developed a method for extracting high-quality gasoline by catalytically 'cracking' low-grade crude oil. In 1930 he emigrated to the USA to advance this 'Houdry process', which soon revolutionized the production of gasoline around the world. During World War II he developed a method for producing synthetic rubber. Holder of over 100 patents, he also founded the Oxy-Catalyst company to research cancer.

He was born in Domont, France.

Kendall, Edward Calvin (1886–1972)

US biochemist who was awarded a Nobel Prize for Physiology or Medicine in 1950, with Philip Hench and Tadeus Reichstein, for their work on the

US biochemist Edward Kendall (left) and US physician Philip Hench, who shared the 1950 Nobel Prize for Physiology or Medicine for their work on hormones. Hulton-Deutsch Collection/CORBIS

structure and biological effects of hormones of the adrenal cortex. In 1914 Kendall isolated the hormone thyroxine, the active compound of the thyroid gland. He went on to work on secretions from the adrenal gland, among which he discovered the steroid cortisone.

Kendall was born in Connecticut and studied at Columbia University, New York. From 1914 he worked at the Mayo Foundation, Minnesota, USA, becoming professor there in 1921.

Hench was a physician interested in arthritis who was familiar with the experience that in some situations, such as during pregnancy, patients with arthritis improved. He and Kendall discussed whether cortisone, which Kendall's work had shown to have important metabolic effects, was involved in these temporary improvements. They discovered by giving a severely incapacitated patient cortisone that it was an effective treatment for rheumatoid arthritis.

Kroto, Harold W(alter) (1939–)

English chemist who with US chemists Robert F Curl and Richard E Smalley shared the Nobel Prize for Chemistry in 1996 for their discovery of fullerenes.

Using spectroscopy, Kroto had discovered the existence of long-chain molecules composed of only carbon and nitrogen in the atmospheres of carbon-rich giant stars and in interstellar gas clouds. He was investigating how these carbon chains formed when he and his associates discovered fullerenes.

Kroto had contacted Smalley at Rice University, Texas, USA, and in 1985, together with Robert Curl, conducted a series of experiments using the laser supersonic cluster beam apparatus built by Smalley. They vaporized graphite in an inert atmosphere of helium and analysed the spectra of the resulting vapour. This analysis showed that they had not only succeeded in generating carbon chain molecules, but had also created a previously unknown form of carbon.

Until then only two stable forms or allotropes of carbon had been known, diamond and graphite. The new allotrope was mainly composed of 60 carbon atoms linked together in 12 pentagons and 20 hexagons that fitted together to form a hollow sphere. Designated as C_{60}, this was shown to be only one of a family of similar compounds, of which C_{70}, C_{76}, C_{78}, and C_{84} have since been identified.

Kroto and his colleagues named this new form of carbon 'fullerene' after the US architect Richard Buckminster Fuller, because of the structural similarity between the carbon molecules and the geodesic dome which Fuller designed. They are also commonly referred to as buckminsterfullerenes or 'buckyballs'.

Fullerene chemistry is now an established branch of organic chemistry and many novel applications are being investigated. Already, new molecules based on a metal ion enclosed by a 'buckyball' cage are being studied. The potential uses of fullerenes are diverse, ranging from lubricants and semiconductors, to a precursor for new drugs.

Kroto was born in Wisbech, England. He studied for a PhD in chemistry at the University of Sheffield, England, which he was awarded in 1964. He joined the faculty of the University of Sussex in 1967 and became a professor of chemistry there in 1985. He became a Royal Society professor in 1991. He was knighted in 1996. *See illustration on page 73.*

Kuhn, Richard (1900–1967)

Austrian-born German chemist who determined the structures of vitamins A, B_2, and B_6 in the 1930s, having isolated them from cow's milk. He was awarded the Nobel Prize for Chemistry in 1938 for his research into carotenoids and vitamins, but was unable to receive it until after World War II.

Kuhn was born in Vienna and educated there and at Munich, Germany. In 1929 he became professor at Heidelberg and director of the Kaiser Wilhelm (later Max Planck) Institute for Medical Research, and in 1937 took over the administration of the institute. He was unable to receive his Nobel prize until 1945.

Kuhn's early research concerned the carotenoids, the fat-soluble yellow pigments found in plants which are precursors of vitamin A.

In the 1940s, Kuhn continued to carry out research on carbohydrates, studying alkaloid glycosides such as those that occur in tomatoes, potatoes, and other plants of the genus *Solanum.* In 1952, he returned to experiments with milk, extracting carbohydrates from thousands of litres of milk using chromatography. This work led in the 1960s to the investigation of similar sugar-type substances in the human brain.

Lehn, Jean-Marie (1939–)

French chemist who demonstrated for the first time how metal ions could be made to exist in a non-planar structure, tightly bound into the cavity of a crown ether molecule, explaining a possible mechanism for the transfer of metal ions across biological membranes. He was rewarded for his efforts when he was awarded the Nobel Prize for Chemistry in 1987 with Charles Pedersen and Donald Cram for their work on molecules with highly selective structure-specific interactions.

Lehn worked on the structures of crown ethers, which had just been discovered by his contemporary Pedersen and were found to have strong affinities for metal ions. Lehn demonstrated that if two nitrogen atoms replaced oxygen atoms in the 'crown', a three-dimensional structure could be made by connecting two crowns – which Lehn called a 'cryptand'.

Lehn was then able to produce cryptands that could bind to molecules such as neurotransmitters. His research expanded the branch of organic chemistry known as host-guest chemistry and introduced the concept of supramolecules.

Lehn was born in Rosheim, France, and went to Strasbourg University. He studied the terpenes, and then at Harvard, Lehn worked on vitamin B_{12} synthesis. He moved between the two universities, gained a professorship, and finally settled at the College de France, Paris, where he held a chair in chemistry from 1979.

Lipmann, Fritz Albert (1899–1986)

German-born US biochemist who was awarded a Nobel Prize for Physiology or Medicine in 1953 for his discovery of coenzyme A, a nonprotein compound that acts in conjunction with enzymes to catalyse metabolic reactions leading up to the Krebs cycle. He investigated the means by which the cell acquires energy and highlighted the crucial role played by the energy-rich phosphate molecule adenosine triphosphate (ATP). He shared the prize with Hans Krebs.

Lipmann was born in Königsberg, Germany. He moved to the USA in 1939.

Molina, Mario (1943–)

Mexican chemist who shared the Nobel Prize for Chemistry in 1995 with Paul Crutzen and F Sherwood Rowland for their work in atmospheric chemistry, particularly concerning the formation and decomposition of ozone. They explained the chemical reactions which are destroying the ozone layer.

The power of nitrogen oxides to decompose ozone was pointed out in 1970 by Crutzen. In 1974 Rowland and Molina published a widely read article on the threat to the ozone layer posed by chlorofluorocarbons (CFCs) used in refrigerators and aerosol cans. They pointed out that CFCs could gradually be carried up into the ozone layer. Here, under the influence of intense ultraviolet light, the CFCs would decompose into their constituents, notably chlorine atoms which decompose ozone in similar ways to nitrogen oxides. They calculated that if the use of CFCs continued at an unaltered rate the ozone layer would be seriously depleted after a few decades. Molina's and Rowland's work led to restrictions on CFC use in the late 1970s and early 1980s.

Molina was born in Mexico City and educated in physical chemistry at the University of California, Berkeley, USA. The first Mexican to receive a Nobel prize for science, he announced in February 1996 that he would donate part of his prize to fund the training of environmental researchers from Latin America and other developing countries. Molina is currently at the Massachusetts Institute of Technology, Cambridge, Massachusetts, USA.

Müller, Paul Hermann (1899–1965)

Swiss chemist who was awarded a Nobel Prize for Physiology or Medicine in 1948 for his discovery of the first synthetic contact insecticide, DDT, in 1939.

Müller was born in Olten, Solothurn, and studied at Basel. He went to work for the chemical firm of J R Geigy, researching principally into dyestuffs and tanning agents; he subsequently joined the staff of Basel University.

In 1935 Müller started the search for a substance that would kill insects quickly, but have little or no poisonous effect on plants and animals, unlike the arsenical compounds then in use. He concentrated his search on chlorine compounds and in 1939 synthesized DDT.

The Swiss government successfully tested DDT against the Colorado potato beetle in 1939 and by 1942 it was in commercial production. Its first important use was in Naples, Italy, where a typhus epidemic in the period 1943–44 was ended when the population was sprayed with DDT to kill the body lice that are the carriers of typhus.

Northrop, John Howard (1891–1987)

US chemist. He was awarded the Nobel Prize for Chemistry in 1946 for his work in the 1930s when he crystallized a number of enzymes, including pepsin and trypsin, showing conclusively that they were proteins. He shared the award with Wendell Stanley and James Sumner.

Northrop was born in Yonkers, New York, and studied at Columbia University. In 1916 he was appointed to the staff of the Rockefeller Institute and remained there for the rest of his career.

Pauling, Linus Carl (1901–1994)

US theoretical chemist and biologist. He was awarded the Nobel Prize for Chemistry in 1954 for his study of the nature of chemical bonds, especially in complex substances. His ideas are fundamental to modern theories of molecular structure. He also investigated the properties and uses of vitamin C as related to human health. He was awarded the Nobel Prize for Peace

US chemist Linus Pauling holds up a sign in 1962 as he pickets the White House during a mass protest against resumption of US atmospheric nuclear testing. Bettmann/CORBIS

in 1962 for having campaigned for the control of nuclear weapons and nuclear testing.

Pauling's work on the nature of the chemical bond included much new information about interatomic distances. Applying his knowledge of molecular structure to proteins in blood, he discovered that many proteins have structures held together with hydrogen bonds, giving them helical shapes.

He was a pioneer in the application of quantum-mechanical principles to the structures of molecules, relating them to interatomic distances and bond angles by X-ray and electron diffraction, magnetic effects, and thermochemical techniques. In 1928, Pauling introduced the concept of hybridization of bonds. This provided a clear, basic insight into the framework structure of all carbon compounds – in effect, of the whole of organic chemistry. He also studied electronegativity of atoms and polarization (location of electrons) in chemical bonds. Electronegativity values can be used to show why certain substances, such as hydrochloric acid, are acid, whereas others, such as sodium hydroxide, are alkaline. Much of this work was consolidated in his book *The Nature of the Chemical Bond* (1939).

In his researches on blood in the 1940s, Pauling investigated immunology and sickle-cell anaemia. Later work confirmed his conviction that the disease is genetic and that normal haemoglobin and the haemoglobin in sickle cells differ in electrical charge. Pauling's work provided a powerful impetus to Crick and Watson in their search for the structure of DNA.

Pauling was born in Portland, Oregon, and studied at Oregon State Agricultural College, getting his PhD from the California Institute of Technology (Caltech). In Europe 1925–27, he met the chief atomic scientists of the day. He became professor at Caltech in 1931, and was director of the Gates and Crellin Laboratories 1936–58 and of the Linus Pauling Institute of Science and Medicine in Menlo Park, California 1973–75.

During the 1950s he became politically active, his special concern being the long-term genetic damage resulting from atmospheric nuclear bomb tests. In this, he came into conflict with the US establishment and several of his science colleagues. He was denounced as a pacifist and a communist, his passport was withdrawn 1952–54, and he was obliged to appear before the US Senate Internal Security Committee. One item in his sustained wide-ranging campaign was his book *No More War!* (1958). He presented to the UN a petition signed by 11,021 scientists from 49 countries urging an end to nuclear-weapons testing, and during the 1960s spent several years on a study of the problems of war and peace at the Center for the Study of Democratic Institutions in Santa Barbara, California.

Pedersen, Charles (1904–1990)

US organic chemist who shared the Nobel Prize for Chemistry in 1987 with Jean Lehn and Donald Cram for his discovery of 'crown ether', a cyclic polyether – a molecule with 12 carbon atoms and six oxygen atoms arranged in a crown-like structure. Its discovery opened up the field of guest-host chemistry.

Pedersen found that crown ether, now part of a class of crown ethers, has unusual properties, such as being able to dissolve sodium hydroxide because it binds alkali metal ions very strongly, forming a complex. This discovery enabled scientists to study metal ion transport across cell membranes.

Pedersen was born in Pusan, Korea. His mother was Japanese and his father was a Norwegian mining engineer, and in the 1920s the family settled in the USA. After studying chemical engineering at the University of Dayton, Ohio, he completed a master's degree in organic chemistry at the Massachusetts Institute of Technology. He spent most of his working life as part of the research team for DuPont, until his retirement in 1969. His discovery came by accident. While working on synthetic rubber, he spotted that one of his preparations had been contaminated with an impurity.

Perey, Marguérite (Catherine) (1909–1975)

French nuclear chemist who discovered the radioactive element francium in 1939. Her career, which began as an assistant to Marie Curie in 1929, culminated with her appointment as professor of nuclear chemistry at the University of Strasbourg in 1949 and director of its Centre for Nuclear Research in 1958.

Perey was born at Villemomble and educated at the Faculté des Sciences in Paris.

Perutz, Max Ferdinand (1914–)

Austrian-born British biochemist who shared the Nobel Prize for Chemistry in 1962 with his co-worker John Kendrew for work on the structure of the haemoglobin molecule.

Perutz, born and educated in Vienna, moved to Britain in 1936 to work on X-ray crystallography at the Cavendish Laboratory, Cambridge. After internment in Canada as an enemy alien during World War II, he returned to Cambridge and in 1947 was appointed head of the new Molecular Biology Unit of the Medical Research Council (MRC). From 1962 he was chair of the MRC's new Laboratory of Molecular Biology.

Austrian-born British biochemist Max Perutz (left) and English biochemist Paul Kendrew at Cambridge University in 1962. Hulton-Deutsch Collection/CORBIS

Perutz first applied the methods of X-ray diffraction to haemoglobin in 1937, but it was not until 1953 that he discovered that if he added a single atom of a heavy metal such as gold or mercury to each molecule of protein, the diffraction pattern was altered slightly. Using this technique, he had worked out the precise structure of haemoglobin by 1960.

Later, Perutz tried to interpret the mechanism by which the haemoglobin molecule transports oxygen in the blood, realizing that an inherited disorder such as sickle-cell anaemia could be caused by a mutation of this molecule.

Plunkett, Roy J (1910–1994)

US chemist and inventor. In 1938, after only two years as a research chemist for E I du Pont de Nemours & Co. he discovered the plastic Teflon, a breakthrough that led to many new fluorochemical products now widely used in the electronics, plastics, and aerospace industries. He remained an effective manager of research, development and production for DuPont until he retired in 1975. Plunkett was born in New Carlisle, Ohio.

Porter, Rodney Robert (1917–1985)

English biochemist who was awarded a Nobel Prize for Physiology or Medicine in 1972 for his work on the chemical structure of antibodies. In 1962 he proposed a structure for human immunoglobulin (antibody) in which the molecule was seen as consisting of four chains.

Porter was born in Liverpool and studied there and at Cambridge. He became professor of immunology at St Mary's Hospital Medical School, London, in 1960. From 1967 he was professor of biochemistry at Oxford University.

Basing his research on the work of US immunologist Karl Landsteiner, Porter studied the structural basis of the biological activities of antibodies, proposing in 1962 a structure for gamma globulin. He also worked on the structure, assembly, and activation mechanisms of the components of a substance known as complement. This is a protein that is normally present in the blood, but disappears from the serum during most antigen–antibody reactions. In addition, Porter investigated the way in which immunoglobulins interact with complement components and with cell surfaces.

Prelog, Vladimir (1906–1998)

Bosnian-born Swiss organic chemist who studied alkaloids and antibiotics. The comprehensive molecular topology that evolved from his work on stereochemistry is gradually replacing classical stereochemistry. He shared the Nobel Prize for Chemistry in 1975 for his work on the stereochemistry of organic molecules and their reactions.

Prelog was born in Sarajevo and studied in Czechoslovakia at the Institute of Technology in Prague. In 1935 he went back to Yugoslavia to lecture at Zagreb, but in 1941, after the German occupation at the beginning of World War II, Prelog moved to the Federal Institute of Technology, Zürich, where he became professor in 1957.

Alkaloids were the subject of Prelog's early research, and he derived the structures of quinine, strychnine, and steroid alkaloids from plants of the genera *Solanum* and *Veratrum*. His studies of lipoid extracts from animal organs also resulted in the discovery of various steroids and the elucidation of their structures. He investigated metabolic products of micro-organisms and with a number of other researchers isolated various new complex natural products that have interesting biological properties. These include antibiotics and bacterial growth factors.

Rowland, F Sherwood (1927–)

US chemist who shared the Nobel Prize for Chemistry in 1995 with Mario Molina and Paul Crutzen for their work in atmospheric chemistry, particularly concerning the formation and decomposition of ozone. They explained the chemical reactions that are destroying the ozone layer.

The power of nitrogen oxides to decompose ozone was pointed out in 1970 by Crutzen. In 1974 Rowland and Molina published a widely read article on the threat to the ozone layer posed by chlorofluorocarbons (CFCs) used in refrigerators and aerosol cans. They pointed out that CFCs could gradually be carried up into the ozone layer. Here, under the influence of intense ultraviolet light, the CFCs would decompose into their constituents, notably chlorine atoms which decompose ozone in similar ways to nitrogen oxides. They calculated that if the use of CFCs continued at an unaltered rate the ozone layer would be seriously depleted after a few decades. Molina's and Rowland's work led to restrictions on CFC use in the late 1970s and early 1980s.

Rowland was born in Delaware, Ohio, USA, and educated in chemistry at the University of Chicago, USA, receiving his doctorate in 1952. He is currently at the University of California at Irvine, California, USA. He was selected to be one of a US delegation of 20 to the World Conference on Science organized by UNESCO and the International Council for Science in 1999.

Sabatier, Paul (1854–1941)

French chemist. He found in 1897 that if a mixture of ethylene and hydrogen was passed over a column of heated nickel, the ethylene changed into ethane. Further work revealed that nickel could be used to catalyse numerous chemical reactions. He was awarded the Nobel Prize for Chemistry in 1912 for finding the method of catalytic hydrogenation of organic compounds.

Sabatier was born in Carcassone, Aude, and studied at the Ecole Normale Supérieure in Paris. From 1884 he was professor at Toulouse.

With his assistant Abbé Jean-Baptiste Senderens (1856–1936), Sabatier extended the nickel-catalyst method to the hydrogenation of other unsaturated and aromatic compounds, and synthesized methane by the hydrogenation of carbon monoxide. He later showed that at higher temperatures the same catalysts can be used for dehydrogenation, enabling him to prepare aldehydes from primary alcohols and ketones from secondary alcohols.

Sabatier later explored the use of oxide catalysts, such as manganese oxide, silica, and alumina. Different catalysts often gave different products from the same starting material.

Alumina, for example, produced olefins (alkenes) with primary alcohols, which yielded aldehydes with a copper catalyst.

Sanger, Frederick (1918–)

English biochemist. He was awarded the Nobel Prize for Chemistry in 1958 for determining the structure of insulin, and again in 1980 for work on the chemical structure of genes. He was the first person to be awarded the Chemistry Prize twice.

Sanger's second Nobel prize was shared with two US scientists, Paul Berg and Walter Gilbert, for establishing methods of determining the sequence of nucleotides strung together along strands of RNA and DNA. He also worked out the structures of various enzymes and other proteins.

Sanger was born in Rendcombe, Gloucestershire, and studied at Cambridge, where he spent his whole career. In 1961 he became head of the Protein Chemistry Division of the Medical Research Council's Molecular Biology Laboratory.

Between 1943 and 1953, Sanger and his co-workers determined the sequence of 51 amino acids in the insulin molecule. By 1945 he had discovered a compound, ***Sanger's reagent*** (2,4-dinitrofluorobenzene), which attaches itself to amino acids, and this enabled him to break the protein chain into smaller pieces and analyse them using paper chromatography.

From the late 1950s, Sanger worked on genetic material, and in 1977 he and his co-workers announced that they had established the sequence of more than 5,000 nucleotides along a strand of RNA from a bacterial virus called R17. They later worked out the order for mitochondrial DNA, which has approximately 17,000 nucleotides.

Seaborg, Glenn Theodore (1912–1999)

US nuclear chemist. He shared the Nobel Prize for Chemistry in 1951 with his co-worker Edwin McMillan for his discovery of plutonium and research on the transuranic elements.

All transuranic elements are radioactive and none occurs to any appreciable extent in nature; they are synthesized by transmutation reactions. Seaborg was involved in the identification of plutonium (atomic number 94) in 1940, americium (95) in 1944–45, curium (96) in 1944, berkelium (97) in 1949, californium (98) in 1950, einsteinium (99) in 1952, fermium (100) in 1952, mendelevium (101) in 1955, and nobelium (102) in 1958.

Seaborg was born in Ishpeming, Michigan, and studied at the University of California. During part of World War II he was at the metallurgical laboratory at Chicago University, where much of the early work on the atomic bomb was carried out. He was professor of chemistry at the University of California at Berkeley 1945–61, and chair of the Atomic Energy Commission 1961–71, encouraging the rapid growth of the US nuclear-power industry.

US nuclear chemist Glenn Seaborg in his laboratory in 1980. The very first samples of plutonium-239 were stored in cigar boxes provided by fellow chemist Gilbert Lewis.
Roger Ressmeyer/CORBIS

For his leadership in the development of nuclear chemistry and atomic energy, he received the 1959 Enrico Fermi award. In 1971 he returned to academic life at Berkeley. He received the Priesley Medal of the American Chemical Society (1979), the Henry De Wolf Smyth Award of the American Nuclear Society (1982), and the Actinide Award (1984). In 1997 element 106 was officially named seaborgium, the first time an element had been named after a living person.

Soddy, Frederick (1877–1956)

English physical chemist who pioneered research into atomic disintegration and coined the term isotope. He was awarded the Nobel Prize for Chemistry in 1921 for investigating the origin and nature of isotopes.

The displacement law, introduced by Soddy in 1913, explains the changes in atomic mass and atomic number for all the radioactive intermediates in the decay processes.

After his chemical discoveries, Soddy spent some 40 years developing a theory of 'energy economics', which he called 'Cartesian economics'. He argued for the abolition of debt and compound interest, the nationalization of credit, and a new theory of value based on the quantity of energy contained in a thing.

Soddy was born in Eastbourne, Sussex, and studied at Oxford. He worked 1900–02 with physicist Ernest Rutherford at McGill University in Montréal, Canada. Soddy was professor at Aberdeen 1914–19 and at Oxford 1919–36. He opposed military use of atomic power.

Soddy and Rutherford postulated that radioactive decay is an atomic or subatomic process, and formulated a disintegration law. They also predicted that helium should be a decay product of radium, a fact that Soddy proved spectrographically in 1903.

His works include *Chemistry of the Radio-Elements* (1912–14), *The Interpretation of the Atom* (1932), and *The Story of Atomic Energy* (1949).

Staudinger, Hermann (1881–1965)

German organic chemist who was awarded the Nobel Prize for Chemistry in 1953 for his discoveries in macromolecular chemistry, of which he was the founder. He carried out pioneering research into the structure of albumen and cellulose.

To measure the high molecular weights of polymers he devised a relationship, now known as ***Staudinger's law,*** between the viscosity of polymer solutions and their molecular weight.

Staudinger was born in Worms, Hesse, and studied at a number of German universities. He became professor in 1908 at the Technische

German-born US biochemist Franz Lipmann (left) talking to German organic chemist Hermann Staudinger in 1955, at the Nobel Prize Scientists for Ban on Atomic War in Lindau, West Germany. Bettmann/CORBIS

Hochschule in Karlsruhe, moved to Zürich, Switzerland, 1912, and from 1926 was at the University of Freiburg im Breisgau, where in 1940 he was made director of the Chemical Laboratory and Research Institute for Macromolecular Chemistry.

He devised a new and simple synthesis of isoprene (the monomer for the production of the synthetic rubber polyisoprene) in 1910.

Most chemists thought that polymers were disorderly conglomerates of small molecules, but from 1922 Staudinger put forward the view that polymers are giant molecules held together with ordinary chemical bonds. To give credence to the theory, he made chemical changes to polymers that left their molecular weights almost unchanged; for example, he hydrogenated rubber to produce a saturated hydrocarbon polymer.

In his book *Macromolekulare Chemie und Biologie* (1947), Staudinger anticipated the molecular biology of the future.

Sumner, James (Batcheller) (1887–1955)

US biochemist. Sumner shared the Nobel Prize for Chemistry in 1946 with John Northrop and Wendell Stanley for his work in 1926 when he succeeded in crystallizing the enzyme urease and demonstrating its protein nature.

He was born in Canton, Massachusetts. He spent his entire career at Cornell 1915–55.

Svedberg, Theodor (1884–1971)

Swedish chemist. In 1923 he constructed the first ultracentrifuge, a machine that allowed the rapid separation of particles by mass. This can reveal the presence of contaminants in a sample of a new protein, or distinguish between various long-chain polymers. He was awarded the Nobel Prize for Chemistry in 1926 for his investigation of dispersed systems.

Svedberg was born near Gävle, studied at Uppsala and spent his career there, as professor 1912–49 and head of the Institute of Nuclear Chemistry 1949–67.

Svedberg prepared a number of new organosols from more than 30 metals. Through an ultramicroscope, he studied the particles in these sols and confirmed Albert Einstein's theories about Brownian movement.

Svedberg discovered that thorium-X (now known as lead-208) crystallizes with lead and barium salts (but not with others), anticipating English chemist Frederick Soddy's demonstration of the existence of isotopes.

Svedberg also investigated, about 1923, the chemistry involved in the formation of latent images in photographic emulsions.

Working on synthetic polymers, Svedberg introduced electron microscopy to study natural and regenerated cellulose, X-ray diffraction techniques to investigate cellulose fibres, and electron diffraction to analyse colloidal micelles and crystallites.

Synge, Richard Laurence Millington (1914–1994)

British biochemist who improved paper chromatography (a means of separating mixtures) to the point where individual amino acids could be identified. He shared the Nobel Prize for Chemistry in 1952 with his colleague Archer Martin for the development in 1944 of the technique known as partition chromatography.

Martin and Synge worked together at Cambridge and at the Wool Industries Research Association in Leeds, Yorkshire. Their chromatographic method became an immediate success, widely adopted. It was soon demonstrated that not only the type but the concentration of each amino acid can be determined.

Synge was born in Liverpool and studied at Winchester College and at Cambridge University. In 1948 he moved to the Rowett Research Institute, Aberdeen, Scotland, where he remained in charge of protein chemistry until 1967. From 1967 until his retirement in 1974 he worked as a biochemist at the Food Research Institute of the Agricultural Research Council in Norwich. He was also honorary professor of biology at the University of East Anglia. Synge was very active in the peace movement.

In the early 1940s there were only crude chromatographic techniques available for separating proteins in reasonably large samples; no method existed for the separation of the amino acids that make up proteins. Martin and Synge developed the technique of paper chromatography, using porous filter paper to separate out amino acids using a solvent. A minute quantity of the amino acid solution is placed near the edge of the filter paper; once dry, it is dipped (or suspended) in a solvent. As the solvent passes the mixture, the amino acids move with it, but they do so at different rates and so become separated. Once dry, the paper is sprayed with a developer and the amino acids show up as coloured dots. Synge and Martin announced their technique in 1944; it was soon being applied to a wide variety of experimental problems.

Paper chromatography is so precise that it can be used to identify amino acid concentration, as well as type, enabling Synge to work out the exact structure of the antibiotic peptide Gramicidin-S, a piece of research important to Frederick Sanger's determination of the structure of insulin in 1953.

Szent-Györgyi, Albert von Nagyrapolt (1893–1986)

Hungarian-born US biochemist who was awarded the Nobel Prize for Physiology or Medicine in 1937 for his investigation of biological oxidation processes and of the action of ascorbic acid (vitamin C).

In 1928 Szent-Györgyi isolated a substance from the adrenal glands that he named hexuronic acid; he also found it in oranges and paprika, and in 1932 proved it to be vitamin C.

Szent-Györgyi also studied the uptake of oxygen by muscle tissue. In 1940 he isolated two kinds of muscle protein, actin and myosin, and named the combined compound actinomyosin. When adenosine triphosphate (ATP) is added to it, a change takes place in the relationship of the two components which results in the contraction of the muscle. When a muscle contracts, myosin and actin move in relation to one another powered by energy released by hydrolysis of ATP and elevated levels of calcium in the muscle cells.

In the 1960s Szent-Györgyi began studying the thymus gland, and isolated several compounds from the thymus that seem to be involved in the control of growth.

Hungarian-born US biochemist Albert Szent-Györgyi in his laboratory in 1955.
Bettmann/CORBIS

Szent-Györgyi was born in Budapest and educated there and at universities elsewhere in Europe and in the USA. He was active in the anti-Nazi underground movement during World War II; after the war he became professor at Budapest. In 1947 he emigrated to the USA, where he became director of the National Institute of Muscle Research at Woods Hole, Massachusetts. From 1975 he was scientific director of the National Foundation for Cancer Research. His last work was *Electronic Biology and Cancer* (1976).

Tiselius, Arne Wilhelm Kaurin (1902–1971)

Swedish chemist who developed a powerful method of chemical analysis known as electrophoresis. He applied his new techniques to the analysis of animal proteins. He was awarded the Nobel Prize for Chemistry in 1948 for his researches in electrophoresis and adsorption analysis, and discoveries concerning serum proteins.

Tiselius was born in Stockholm and studied at Uppsala, where he spent his career. From 1938 he was director of the Institute of Biochemistry.

Working at Princeton in the USA 1934–35, Tiselius investigated zeolite minerals and their capacity to retain their crystal structure by exchanging their water of crystallization for other substances. The crystal structure remains intact even after the water has been removed under vacuum. He studied the optical changes that occur when the dried crystals are rehydrated.

Tiselius first used electrophoresis in the 1920s. In the 1930s he separated the proteins in horse serum and revealed for the first time the existence of three components which he named α-, β-, and γ-globulin. Later he developed new techniques in chromatography.

Tiselius founded the Nobel Symposia, which take place every year in each of the five prize fields to discuss the social, ethical, and other implications of the award-winning work.

Urey, Harold Clayton (1893–1981)

US chemist. In 1932 he isolated heavy water and was awarded the Nobel Prize for Chemistry in 1934 for his discovery of deuterium (heavy hydrogen).

During World War II he was a member of the Manhattan Project, which produced the atomic bomb, and after the war he worked on tritium (another isotope of hydrogen, of mass 3) for use in the hydrogen bomb, but later he advocated nuclear disarmament and world government.

Urey was born in Indiana and educated at Montana State University. He became professor of chemistry at Columbia 1934, and was at Chicago 1945–58.

After deuterium, Urey went on to isolate heavy isotopes of carbon, nitrogen, oxygen, and sulphur. His group provided the basic information for the separation of the fissionable isotope uranium-235 from the much more common uranium-238.

Urey also developed theories about the formation of the Earth. He thought that the Earth had not been molten at the time when its materials accumulated. In 1952, he suggested that molecules found in its primitive atmosphere could have united spontaneously to give rise to life. The Moon, he believed, had a separate origin from that of the Earth.

Wittig, Georg (1897–1987)

German chemist who shared the Nobel Prize for Chemistry in 1979 for his method of synthesizing olefins (alkenes) from carbonyl compounds. This reaction is often termed the ***Wittig synthesis.***

Wittig was born in Berlin and studied at Kassel and Marburg. He was professor at Freiburg 1937–44, Tübingen 1944–56, and Heidelberg 1956–67.

In the Wittig reaction, which he first demonstrated in 1954, a carbonyl compound (aldehyde or ketone) reacts with an organic phosphorus compound, an alkylidenetriphenylphosphorane, $(C_6H_5)_3P{=}CR_2$, where R is a hydrogen atom or an organic radical. The alkylidene group ($=CR_2$) of the reagent reacts with the oxygen atom of the carbonyl group to form a hydrocarbon with a double bond, an olefin (alkene). In general:

$$(C_6H_5)_3P{=}CR_2 + R_2'CO \quad (C_6H_5)_3PO + R_2C{=}CR_2$$

The reaction is widely used in organic synthesis, for example to make squalene (the synthetic precursor of cholesterol) and vitamin D_3.

Woodward, Robert Burns (1917–1979)

US chemist who worked on synthesizing a large number of complex molecules. These included quinine in 1944, cholesterol in 1951, chlorophyll in 1960, and vitamin B_{12} in 1971. He was awarded the Nobel Prize for Chemistry in 1965 for his work in organic synthesis.

Woodward was born in Boston and studied at the Massachusetts Institute of Technology. He worked throughout his career at Harvard, becoming professor in 1950.

In 1947 Woodward worked out the structure of penicillin and two years later that of strychnine. In the early 1950s, he began to synthesize steroids, and in 1954 he synthesized the poisonous alkaloid strychnine and lysergic acid, the basis of the hallucinogenic drug LSD. In 1956 he made reserpine, the first of the tranquillizing drugs. Turning his attention again to antibiotics, he and his co-workers produced a tetracycline in 1962 and cephalosporin C in 1965. The synthesis of vitamin B_{12} was made in collaboration with Swiss chemists and took ten years.

Ziegler, Karl (1898–1973)

German organic chemist. He shared the Nobel Prize for Chemistry in 1963 with Giulio Natta of Italy for his work on the chemistry and technology of large polymers. He combined simple molecules of the gas ethylene (ethene) into the long-chain plastic polyethylene (polyethene).

Ziegler and Natta discovered in 1953 a family of stereo-specific catalysts capable of introducing an exact and regular structure to various polymers. This discovery formed the basis of nearly all later developments in synthetic plastics, fibres, rubbers, and films derived from such olefins as ethylene (ethene) and butadiene (but-1,2:3,4-diene).

Ziegler was born near Kassel and studied at Marburg. From 1943 he was director of the Kaiser Wilhelm (later Max Planck) Institute for Coal Research in Mülheim.

In 1933, Ziegler discovered a method of making compounds that contain large rings of carbon atoms. Later he carried out research on the organic compounds of aluminium. Using electrochemical techniques, he prepared various other metal alkyls from the aluminium ones, including tetraethyl lead, which was used as an additive to petrol.

In 1953 Ziegler found that organometallic compounds mixed with certain heavy metals polymerize ethylene at atmospheric pressure to produce a linear polymer of high molecular weight (relative molecular mass) and with valuable properties, such as high melting point.

Zsigmondy, Richard Adolf (1865–1929)

Austrian-born German chemist who devised and built an ultramicroscope in 1903. The microscope's illumination was placed at right angles to the axis. (In a conventional microscope the light source is placed parallel to the instrument's axis.) Zsigmondy's arrangement made it possible to observe particles with a diameter of 10-millionth of a millimetre. He was awarded the Nobel Prize for Chemistry in 1925 for the elucidation of heterogeneity of colloids.

Zsigmondy was born in Vienna and studied at Munich. He stayed in Germany, becoming professor at Göttingen in 1908. Working at the Glass Manufacturing Company in Jena 1897–1900, Zsigmondy became concerned with coloured and turbid glasses and he invented a type of milk glass. This aroused his interest in colloids, because it is colloidal inclusions that give glass its colour or opacity. His belief that the suspended particles in gold sols are kept apart by electric charges was generally accepted, and the sols became model systems for much of his later work on colloids.

Using the ultramicroscope Zsigmondy was able to count the number of particles in a given volume and indirectly estimate their sizes. He showed that colour changes in sols reflect changes in particle size caused by coagulation when salts are added, and that the addition of agents such as gelatin stabilizes the colloid by inhibiting coagulation.

Part Two

4 Directory of Organizations and Institutions

The American Chemical Society

The world's largest professional organization for chemists, with over 155,000 members. The organization publishes thousands of scientific papers and articles each year and holds annual conferences, where chemical professionals can meet and discuss their work. It provides career advice, offers courses in chemistry at all levels of expertise, and keeps its members informed as to the latest international and political developments in science.

Address
American Chemical Society
1155 16H Street NW
Washington, DC 20036, USA
phone: +1 (202) 872 4600
Web site: http://www.acs.org

The American Petroleum Institute (API)

A trade association representing the entire petroleum industry of the USA. It originated in 1919 when there was a need to standardize drilling and production equipment, and has developed into the official voice of the industry. The organization provides information regarding all aspects of the industry and its products, contributing over 200,000 publications each year to further knowledge in this field. API organizes training and education and has developed over 500 equipment and operation standards, which are used throughout the world to provide a high level of competence and safety in the industry.

Address
The American Petroleum Institute
1220 L Street NW
Washington, DC, 20005 4070, USA
phone: +1 (202) 682 8000
Web site: http://www.api.org

Argonne National Laboratory Chemistry Division

The chemistry division of an organization founded in 1946 to explore the peaceful uses of atomic energy. The division is largely concerned with solving energy-related problems, and is mainly funded by Basic Energy Sciences at the Department of Energy. The scope of research carried out at Argonne is wide ranging: as well as nuclear chemistry, there are programmes studying organic superconductivity, petroleum chemistry, photosynthesis, and metal cluster chemistry.

Address
Argonne National Laboratory
9700 S Cass Avenue
Argonne, IL 60449, USA
phone: +33 (630) 252 2000
Web site: http://chemistry.anl.gov/

Bayer

An international chemicals and healthcare company. Bayer employs over 143,000 people throughout the world and has annual sales in the region of DM 55 billion. It maintains its market position by virtue of its outstanding commitment to research and development. Vital to this effort is Bayer's Central Research Division, where over 2,000 researchers develop the company's products of the future. The company also maintains close links with academic research in universities and institutions worldwide, and increasingly their research is a product of this cooperation. The total number of research staff at Bayer is over 14,000 and the company spends in the order of DM 4 billion on research and development.

Address
Bayer Corporation
Corporate Communications
100 Bayer Road
Building 4
Pittsburgh, PA 15205 9741, USA
phone: +1 (412) 777 2000
Web site: http://www.bayerus.com

California Institute of Technology Division of Chemistry and Chemical Engineering

Institute concerned with scientific and technological research. California Institute of Technology (Caltech) has a long-held and illustrious reputation over a wide range of research, including chemical. The Division of Chemistry and Chemical Engineering forms one of the six academic divisions at Caltech. Research provides the main emphasis of the graduate programme and is geared towards preparing students for careers in science and science-related fields.

Address
California Institute of Technology
1200 East California Boulevard
Pasadena, CA, 91125, USA
phone: +1 (626) 395-6811
Web site: http://www.caltech.edu/subpages/chem.html

Chemical and Analytical Sciences Division of Oak Ridge National Laboratory

Part of the Advanced Materials, Physical and Neutron Sciences Directorate, carrying out both fundamental and applied research. The division specializes in analytical chemistry, radiochemistry, geochemistry, and materials chemistry, the majority of the research being funded by the Department of Energy. The Division welcomes weekly tours of the facility both by the general public and by visiting researchers. A key goal of the group is to promote research in restoring and protecting the environment and developing clean sources of energy production.

Address
Oak Ridge National Laboratory
Chemical and Analytical Sciences Division
Oak Ridge, TN 37831, USA
phone: +1 (423) 574 4986
fax: +1 (423) 574 4902
e-mail: vjsornl.gov
Web site: http://www.ornl.gov

Chemical Science and Technology Division of Los Alamos National Laboratory (CST)

One of the largest public chemistry research organizations in the USA. Research in the division is very broad and covers subjects as diverse as environmental chemistry, nuclear chemistry, biochemistry, and reaction chemistry. The main goal of the organization is to promote the development of pure and applied research to solve the complex problems that science faces today, such as energy production, environmental issues, and biotechnology. The division has access to world-class facilities, which provide safe working environments for complex operations such as plutonium chemistry and medical isotope synthesis.

Address
Chemical Science and Technology Division
MS J515 Los Alamos National Laboratory
Los Alamos, NM 87545, USA
phone: +1 (505) 667 4457
Web site: http://aha-public.lanl.gov

Chemical Sciences Division of the Lawrence Berkeley National Laboratory (CSD)

One of Berkeley Laboratory's divisions, which carries out research in the areas of chemical physics, catalysis, surface chemistry, and actinide element chemistry. The effects of actinide elements on the environment are also studied at CSD. Berkeley Laboratory is operated by the University of California as part of the Department of Energy's national laboratory system.

Address
Chemical Sciences Division
Lawrence Berkeley National Laboratory
1 Cyclotron Road
Berkeley, CA 94720, USA
phone: +1 (510) 486 4613
Web site: http://www.lbl.gov/

Chemistry and Materials Division of the Lawrence Livermore National Laboratory

A division which provides chemistry support to the research programmes being carried out at the Lawrence Livermore National Laboratory (LLNL). This research is mainly concerned with aspects of physical chemistry and chemical engineering, although their Polymeric Materials Section provides expertise in polymer synthesis and a range of related subjects. As part of the LLNL, the division is operated by the University of Cali0 fornia.

Address
University of California
Lawrence Livermore National Laboratory
7000 East Avenue
PO Box 808, Livermore
CA 94550, USA
phone: +1 (925) 422 1100
fax: +1 (925) 422 1370
Web site: http://www.llnl.gov

The Dow Chemical Company

The world's fifth largest chemical company. Dow provides plastics, energy, agricultural products, chemicals, plastics, and consumer goods, with annual sales in excess of $20 billion. A true multinational, the company employs over 42,000 people in 164 countries.

Address

Dow Chemical Customer Information Group
Dow Ashman Center
4520 Ashman Street
PO Box 1206
Midland, MI 48642, USA
phone: +1 (800) 258 2436
fax: 517 832 1465
Web site: http://www.dow.com

DuPont

The 15th largest US industrial/service corporation. The company has a distinguished record in research of new materials and developed the plastics Nylon, Teflon, and Lycra. In 2000 it had revenues of $28.3 billion. The company has a wide range of interests, including oil production and refining and the manufacture of petroleum products such as plastics and chemicals. The company also has an extensive research capability, with more than 40 research laboratories throughout the world.

Address

E I du Pont de Nemours and Company
1007 Market Street
Wilmington, DE 19898, USA
phone: +1 (800) 441 7515
e-mail: infodupont.com
Web site: www.dupont.com/index.html

GlaxoWellcome

One the world's leading pharmaceutical companies, being the second largest in both the USA and in Europe. The company is one of the most active industrial contributors to development in the field of biomedical research, employing over 10,000 researchers worldwide. Their research interests encompass the whole of the biomedical field, including genetics, drug development, and biotechnology. The company is also a generous supporter of health-related research, donating $40 million annually.

Address
GlaxoWellcome Inc
P O Box 13398
Research Triangle Park
NC 27709, USA
phone: +1 (919) 248 2100
fax: +1 (919) 248 2381
Web site: http://www.glaxowellcome.co.uk

ICI (Imperial Chemical Industries)

One of the world's largest coatings, speciality chemicals, and materials companies. ICI was formed in 1926 by the merger of Britain's four largest chemical companies. It has built on that sound foundation and now manufactures over 10,000 products in more than 55 countries. Research has always played a key role in ICI's development, and it has an active graduate recruitment programme. The company invests over £270 million annually in research and technology.

Address
Imperial Chemical Industries Plc
9 Millbank
London, SW1P 3JF, UK
phone: +44 (020) 7834 4444
fax: +44 (020) 7834 2042
Web site: http://www.ici.com

Massachusetts Institute of Technology Department of Chemistry

One of the USA's leading chemistry departments. It has a fine tradition of pioneering advances in chemistry and chemical engineering, as well as providing strong leadership in the development of the teaching of chemistry. The department teaches subjects across the full breadth of chemistry as well as providing extra attention to important specialized areas such as organometallic chemistry and biophysical chemistry.

Address
Massachusetts Institute of Technology
Department of Chemistry
77 Massachusetts Avenue
Cambridge, MA 02139-4307, USA
phone: +1 (617) 253 1000
Web site: http://web.mit.edu/chemistry/chem-home.html

Merck

A major international chemical company specializing in pharmaceuticals and laboratory chemicals. It employs 29,000 workers at over 170 companies in 46 countries worldwide. Merck has many research activities involving the development of new products, especially new drugs, environmental analysis, microbiology, and liquid chromatography. The company also invests in the study and implementation of techniques concerning the disposal of chemical waste.

Address
Merck KGaA
Frankfurter Str. 250
D-64293 Darmstadt
Germany
phone: +49 (61) 51 720
fax: +49 (61) 51 72 2000
e-mail: servicemerck.de
Web site: http://www.merck.de

The National Academy of Sciences

An academy established in 1863 at the height of the American Civil War to provide the government with information and advice regarding every aspect of science. The organization has since expanded to include the National Research Council, The National Academy of Engineering, and the Institute of Medicine. The aim of the organization is to provide independent advice on matters of science, technology, and medicine. The institute covers the whole spectrum of scientific issues including the uses of chemicals for agriculture, and air and water pollution.

Address
The National Academy of Sciences
2101 Constitution Avenue NW
Washington, DC 20418, USA
phone: +1 (202) 334 2000
Web site: http://www.nas.edu

New Scientist

British magazine that is read by scientists and all those with an interest in science. It covers new developments from all the scientific disciplines. It is written by scientists in enough detail to interest the professional, but in language also geared towards entertaining the general reader. The Web site includes letters, the latest scientific news, book reviews, special features and an archive of published articles.

Address
New Scientist
Reed Business Information Limited
151 Wardour Street
London, W1V 4BN, UK
phone: +44 (020) 7331 2701
fax: +44 (020) 7331 2777
e-mail: feedbacknewscientist.com
Web site: http://www.newscientist.com

The Nobel Foundation

A private institution whose aim is to recognize and reward novel research. The Nobel Foundation was started in 1900 on the basis of the will of Swedish scientist and industrialist Alfred Nobel. Nobel left a considerable amount of money and assets for the sole purpose of rewarding novel research. The grants awarded with the Nobel prizes help to fund this research. The task of the Foundation is to manage the assets left in Nobel's will to ensure that the committees who award the Nobel prizes are financially independent to chose the most deserving prizewinners.

Address
The Nobel Foundation
Sturegatab 14
Stockholm
Sweden
e-mail: infonobel.se
Web site: http://www.nobel.se

The Royal Australian Chemical Institute

An institute founded in 1917 both as a qualifying body for professional chemists and as an organization committed to the promotion and practice of chemistry. The Institute publishes a monthly magazine, *Chemistry in Australia*, in which research articles, awards, conferences, and scientific reviews are discussed. It has branches that cover a diverse range of specialized disciplines such as cereal chemistry, chemical education, and industrial chemistry. The society also organizes annual events that focus public and professional attention on research developments, notably the National Chemistry Week.

Address
The Royal Australian Chemical Institute
1/21 Vale Street
North Melbourne
Victoria 3051, Australia

phone: +61 (03) 9328 2033
fax: +61 (03) 9328 2670
e-mail: memberraci.org.au
Web site: http://www.raci.org.au

The Royal Society

The oldest professional organization for scientists, founded in 1660. Essentially, it is an independent academy that promotes the natural and applied sciences both on a national and international stage. The society provides a broad range of services to all scientists regardless of their field of expertise and rewards excellence by awarding internationally recognized fellowships, awards, and prizes. The advancement of science is promoted in a number of ways, including lectures, exhibitions, and publications.

Address
The Royal Society
6 Carlton House Terrace
London, SW1Y 5AG, UK
phone: +44 (020) 7839 5561
fax: +44 (020) 7930 2170
Web site: http://www.royalsoc.ac.uk

The Royal Society of Chemistry

The largest professional organization for chemists in Britain, with over 46,000 members. The society provides its members with publishing facilities, an information centre, a society library, and contact information regarding scientists working in the chemical community. It also supports the teaching of chemistry at all levels and is a leading source of chemistry-related information to the general public. The society recognizes advances in the field by the presentation of awards, medals, and lectureships.

Address
The Royal Society of Chemistry
Burlington House
Piccadilly
London, W1J 0BA, UK
phone: +44 (020) 7437 8656
fax: +44 (020) 7287 9825
e-mail: educationrsc.org
Web site: http://www.rsc.org

The Royal Swedish Academy of Sciences

An independent organization, founded in 1739, whose major aim is to promote research in the natural sciences and in mathematics. It actively pursues this aim by publishing research, distributing scientific information, and encouraging contacts between scientist and society. Prizes and grants to further research are provided by the Academy annually from funds held in trust for this purpose. It is this organization which awards the coveted Nobel prizes in chemistry and physics.

Address

The Royal Swedish Academy of Sciences
PO Box 50005
SE-104 05 Stockholm
Sweden
phone: +46 (8) 673 95 00
fax: +46 (8) 15 56 70
e-mail: rsaskva.se
Web site: http://www.kva.se

Shell Chemicals

One of the leading US oil, gas, and petroleum companies. The Shell Chemical Company was founded in the USA in 1929 and pioneered the development of chemicals from natural and refinery gases. The main company business is backed by research and technical services. The company has seven main research facilities that cover subjects such as energy conservation, plastics development, and decreasing waste from chemical processing.

Address

Shell Chemical Company
PO Box 2463
Houston, TX 77252-2463, USA
Web site: http://www.shellchemical.com

Society of the Plastics Industry

The trade association for the plastics industry. Founded in 1937, the aim of the society is to promote the continuing development of plastics while informing the public on the contributions that the industry provides.

Address

The Society of the Plastics Industry
Incorporated
1801 K Street
NW, Suite 600K

Washington, DC 20006, USA
phone: +1 (202) 974 5200
fax: +1 (202) 296 7005
e-mail: feedbacksocplas.org
Web site: www.socplas.org

Tohoku National Industrial Research Institute or TNIRI

An institute established in 1971 with the aim of promoting industrial research in the Tohoku district of Japan. Since then, projects concerning such diverse subjects as geothermal energy and advanced materials have been carried out at the institute. The molecular chemistry division of the institute is dedicated to chemistry-related industrial research.

Address

Tohoku National Industrial Research Institute
4-2-1 Nigatake, Miyagino-ku
Sendai, 983, Japan
phone: +81 (22) 237 5211
fax: +81 (22) 236 6839
Web site: http://www.jst.go.jp/kikan_dir/JN146-f.html

5 Selected Works for Further Reading

Alexander, W and Street, A *Metals in the Service of Man,* 1972

How metals are obtained and worked and the part they play in modern life. An ideal introduction for the general reader.

Asimov, Isaac *Asimov on Chemistry,* 1975

In 17 wide-ranging essays this experienced science popularizer effectively covers the entire field of chemistry in an accessible manner. Everything from the chemistry of the planet Earth, inorganic, organic, and nuclear chemistry are entertainingly treated, and there is an essay on the Nobel prizewinners at the end.

Baldwin, E

A book that first highlighted the fact that biochemistry is an interesting and a quite different discipline from chemistry. Many thousands of undergraduates must have cut their novice teeth on this particular book and were probably grateful.

Calder, Richie *The Life Savers,* 1961

The cover blurb says: 'the enthralling story of today's revolution in medicine – the discovery and development of the life-saving drugs'.

Cousins, E A and Yarsley, V E *Plastics in the Service of Man,* 1956

A description of the structure, manufacture, properties, and the contemporary use of plastics.

Crawford, Elisabeth *Arrhenius: From Ionic Theory to the Greenhouse Effect,* 1997

Thorough account of the scientific interests of Arrhenius, aimed at the student and professional.

Emsley, John *The Consumer's Good Chemical Guide,* 1994

A good guide, thoroughly dippable – just what its title implies.

Emsley, John *The Shocking Story of Phosphorus*, 2000

The history of a fascinating element from the alchemists to the present day.

Findlay, Alexander *Chemistry in the Service of Man*, 1925

A wonderful text from what now seems like a bygone era. Nonetheless it is a substantial introduction to chemistry, very readable, and the chapter on radioactivity and atomic structure, written in the pre-nuclear age, is particularly absorbing.

George Philip Limited *Science & Technology*, 1999

The people, dates, and events in chemistry compared with other sciences and scientists.

Gratzer, W B *Principles of Biochemistry*, 1993

Selected from *Nature*, these short essays were aimed at the working scientist and chart the progress of the then emergent branch of science of molecular biology.

Harrison, Kenneth *A Guide Book to Biochemistry*, 1959

Many standard books on biochemistry are very thick with between 500 and 1,000 pages. This excellent book is a deliberate exception and provides a good guide that is brief and to the point.

Hutton, Kenneth *Chemistry: The Conquest of Materials*, 1957

A good account, by a teacher and writer, of the scope of modern chemistry, telling a clear story from the elements to modern drugs. On the way fuels, modern materials such as plastics, pesticides, and explosives are dealt with at a usefully informed level.

Lehninger, A L *Dynamic Aspects of Biochemistry*, 1967

A comprehensive US undergraduate text, written with masterly clarity. The author properly declares that biochemistry is now the *lingua franca* of the life sciences, and no one consulting this introductory text should be the poorer for having done so.

Nicolaou, K C and Sorensen, E J *Classics in Total Synthesis*, 1996

Description and analysis of the syntheses of milestone molecules including vitamin B_{12}, progesterone, and taxol.

Open University Press *Science and Technology*, 1993

A good work of reference which covers just what its title implies and includes applied chemistry. Well illustrated and up to date, it is a good volume for browsing.

Owen, David *Hidden Evidence,* 2000

The story of forensic science from poison detection to DNA fingerprinting.

Partington, J R *A Short History of Chemistry,* 1937

A prolific author of textbooks on chemistry and author of a multivolume history. This is a clear and authoritative account of the history of chemistry. It draws on many original sources and although it starts with alchemy and ends with radioactivity and the transmutation of the elements, this excellent and concise book concentrates on the foundation of modern chemistry and the great scientists who lead the way.

Pauling, Linus Carl *The Nature of the Chemical Bond and the Structure of Molecules and Crystals: An Introduction to Modern Structural Chemistry,* 1939

A classic by a double Nobel prizewinner, perhaps ambitiously included in a general reading list, but showing how real advances in understanding are made as a result of modern chemical research.

Pyke, Magnus *Butter Side Up or The Delights of Science,* 1978

Reviewed by the *Evening Standard*: 'An opencast mine of unexpected information that can be understood even by someone who cannot tell a Bunsen burner from a laser beam'.

Richardson, R A, Blizzard, A C, and Humphreys, D A *Structure and Change – An Introduction to Chemical Reactions,* 1981

A high-school text that successfully blends a factual and theoretical approach to chemistry. It succeeds in giving an appreciation of the vital role of chemistry in the world.

Royal Society for Chemistry *The Age of the Molecule,* 1999

Highly illustrated and readable look at how molecular research has changed our lives.

Sharp, D W A *The Penguin Dictionary of Chemistry,* 1990

Useful, updated compendium of definitions of chemical terms. Not just examination fodder.

Strathern, Paul *Mendeleyev's Dream,* 2000

The story of the periodic table and how Mendeleyev succeeded in discovering an underlying order to the elements even to the extent of leaving gaps for elements yet to be discovered.

Theiler, Carl R *Men and Molecules*, 1960

What chemistry is and what it does. The author takes the reader through the transformation of first ideas into reality – the building up of mighty industries and the production of astonishing new materials. The reader grows familiar with chemical formulae and their strange names in this well-illustrated book, which also has a very good glossary.

Wills, Christine and Wills, Martin *Organic Synthesis*, 1995

A gentle introduction with examples.

6 Web Sites

Acid and Base pH Tutorial

http://www.science.ubc.ca/~chem/tutorials/pH/launch.html

Aimed at university students with no knowledge of this branch of chemistry, this well-designed tutorial takes you through the basics of acid-base chemistry, including interactive exercises to test your knowledge.

Acids

http://www.purchon.co.uk/science/acids.htm

Offers a clear explanation of the nature and functions of acids. A general definition is followed by more detailed descriptions of some common acids.

Alfred Nobel – His Life and Work

http://www.nobel.se/nobel/alfred-nobel/biographical/life-work/index.html

Presentation of the life and work of Alfred Nobel. The site includes references to Nobel's life in Paris, as well as his frequent travels; his industrial occupations; his scientific discoveries; and especially his work on explosives, which led to the patenting of dynamite. Also covered are his numerous chemical inventions, which included materials such as synthetic leather and artificial silk, his interest in literature and in social and peace-related issues, and of course the Nobel prizes that came as a natural extension of his lifelong interests.

Alkali Metals + H_2O Chemistry

http://www.chem.ualberta.ca/courses/plambeck/p102/p0213x.htm

Good revision aid on the alkali metals and their reaction with water. The reaction of each element is expressed in the form of an equation, then a short section describes the increasing quantity of energy released by each element.

Amino Acids

http://www.chemie.fu-berlin.de/chemistry/bio/amino-acids_en.html

Small but interesting site giving the names and chemical structures of all the common amino acids. The information is available in both English and German.

Analytical Chemistry Basics
http://www.chem.vt.edu/chem-ed/ac-basic.html

Detailed online course, designed for those at undergraduate level, that provides the user with an introduction to some of the fundamental concepts and methods of analytical chemistry. Some of the sections included are gravimetric analysis, titration, and spectroscopy.

Brønsted on Acids and Bases
http://dbhs.wvusd.k12.ca.us/Chem-History/Bronsted-Article.html

Transcript of an article by Brønsted on the properties of acids and bases. The transcript as featured here is not complete, but does cover the essential points in which Brønsted defines the acid and the base. Several examples of his ideal reactions are shown, as well as a brief summary of the properties of the substances.

Carbon Cycle
http://windows.ivv.nasa.gov/cgi-bin/tour_def?link=/earth/Water/co2_cycle.html&frp=/windows3.html&fr=f&sw=false&edu=mid&cdp=/windows3.html&cd=false

Part of a larger site 'Windows to the Universe', this site contains three levels of explanation of the carbon cycle. The site assumes you are an 'Intermediate' student, but you can change this to 'Beginner' or 'Advanced'.

Catalysts
http://www.purchon.co.uk/science/catalyst.html

Clear explanation of catalysts and what they do. Designed to help GCSE/high school students with their revision, this page is part of a larger site of scientific revision aids. The clearly written text includes definitions of important terms such as absorption and intermediate compounds.

Chem101
http://library.thinkquest.org/3310/

Student-created site for those studying chemistry. Aimed at keeping the whole experience fun, this site includes an interactive textbook that covers the basics of the science.

Chemical Bonding
http://edie.cprost.sfu.ca/~rhlogan/bonding.html

Good, text-only introduction to the subject of chemical bonding. A page of general comments is linked to more specific information on ionic and covalent bonding.

Chemical Carousel: A Trip Around the Carbon Cycle
http://library.thinkquest.org/11226/

Captain Carbon invites you to explore the scientific process of the carbon cycle and explains the role of carbon in everyday life.

Chemical Elements.com
http://www.chemicalelements.com/

Searchable, interactive version of the periodic table. This site provides clear information, pictures of atomic structure, and facts about each element. Linked page contains advertising.

Chemical Kinetics: Terms and Concepts
http://www.chem.ualberta.ca/courses/plambeck/p102/p0213x.htm

University of Alberta introduction to chemical kinetics. This well-organized page considers the extent of a chemical reaction, the rate of a chemical reaction, as well as orders and rate constants.

Chemical Reactions
http://www.cornwallis.kent.sch.uk/intranet/subjects/science/chemreact/Index.html

Part of the Web site of the Cornwallis School in Maidstone, Kent, UK, this is a well-organized set of virtual lessons on chemical reactions. Subjects covered include exothermic reactions, speed of reactions, combustion, and air pollution. Each page contains suggested experiments to test the user's knowledge. Linked page contains advertising.

Chemical Stuff
http://www.carlton.paschools.pa.sk.ca/chemical/

School-focused chemistry Web site, providing lively discussion of concepts such as 'The Nature of Proof', a list of experiments on topics such as the nature of gases, and chemistry calculations as well as a links page.

Chemistry Comes Alive!
http://jchemed.chem.wisc.edu/JCESoft/CCA/W1MAIN/CD1W0.HTM

Online version of a CD-ROM designed to accompany the general study of chemistry, including text, videos, and pictures to support the text.

Chemistry Exam Exchange Centre
http://chem.lapeer.org/Exams/Index.html

Approximately 40 chemistry exams and quizzes are freely available to download in PDF format from this site.

Chemistry for Children

http://www.chem4kids.com/

Fun introduction to chemistry. You can take a quiz to test your knowledge of the basics, take a tour around the site, or search for specific information.

Chemistry Functions

http://www.stanford.edu/~glassman/chem/index.htm

Calculate molar conversions online via this handy page. The site also features an interactive periodic table.

Chemistry of Carbon

http://cwis.nyu.edu/pages/mathmol/modules/carbon/carbon1.html

Introduction to carbon, the element at the heart of life as we know it. This site is illustrated throughout and explains the three basic forms of carbon and the importance of the way scientists choose to represent these various structures.

Chemistry Study Cards

http://spusd.k12.ca.us/sphs/science/stdycrds.html

At this site you can download Adobe Acrobat PDF files containing a teacher's study cards to accompany all general chemistry topics.

Chemistry World

http://www.geocities.com/CapeCanaveral/Hangar/6650/

Make your own GAK, a 'silly putty'-style polymer, from the instructions at this fun chemistry site. Light-hearted in approach, the site contains further information on polymers and a useful page on lab safety.

ChemTeam

http://dbhs.wvusd.k12.ca.us/ChemTeamIndex.html

Study resources for high-school chemistry. The ChemTeam site is based around a clickable table of commonly taught high-school chemistry topics, ranging from the history of atomic structure to kinetic molecular theory, stoichometry, thermochemistry, and even some chemical humour to lighten your revision load. There are also links to other helpful sites for chemistry students, including a teacher who will answer chemistry queries online.

ChemTutor

http://www.chemtutor.com/

Beginning with the fundamentals, and graduating to 'the parts of primary chemistry that have been . . . the hardest for students to grasp', this site is a useful resource for any chemistry student.

CHEMystery

http://library.thinkquest.org/3659/

Detailed interactive textbook that includes subjects such as 'Atomic structure and bonding', 'Equilibrium', and 'Thermodynamics'. Use the left-hand menu to navigate to a relevant chapter. Linked page contains advertising.

Chlorine

http://c3.org/

Designed to promote better understanding of the science of chlorine chemistry, examining how it contributes to an enhancement of our standard of living and quality of life.

Chromatography

http://www.eng.rpi.edu/dept/chem=eng/Biotech=Environ/CHROMO/chromintro.html

Explanation of the theory and practice of chromatography. Designed for school students (and introduced by a Biotech Bunny), the contents include equipment, analysing a chromatogram, and details of the various kinds of chromatography.

Classifying Reactions to help with Predicting Reactions

http://www.geocities.com/CapeCanaveral/4724/preactions.html

Revision aid for US students preparing for the AP chemistry test, but of general interest to secondary-school chemists. This page defines five types of chemical reaction: metathesis reactions, redox reactions, organic reactions, complex ion formation, and Lewis acid-base reactions.

Common Sense Science

http://www.commonsensescience.org/

Alternative science site that explores new, 'common sense' models for elementary particles, the atom, and the nucleus.

Computer Assisted Learning in Chemistry

http://members.aol.com/ChangChem/

This chemistry site has an extensive range of online activities, some with accompanying downloadable worksheets. Linked page contains advertising.

Crystallography and Mineralogy

http://www.iumsc.indiana.edu/crystmin.html

Understand the shapes and symmetries of crystallography, with these interactive drawings of cubic, tetrahedral, octahedral, and dodecahedral solids (just drag your mouse over the figures to rotate them).

Definition of Acids and Bases and the Role of Water
http://chemed.chem.purdue.edu/genchem/topicreview/bp/ch11/acidbaseframe.html

This well-organized page contains a table of various definitions of acids and bases. Part of a much larger introductory chemistry site aimed at university students, it shows how scientific knowledge about acids has expanded from the days of Boyle.

Dr Linus Pauling Profile
http://www.achievement.org/autodoc/page/pau0pro-1

Description of the life and works of the multiple Nobel prizewinner Linus Pauling. Pauling is one of the very few people to have won more than one Nobel prize, having won the prize for chemistry in 1954, followed by the prize for peace in 1962. The Web site contains not only a profile and biographical information, but also holds a lengthy interview with Pauling from 1990 accompanied by a large number of photographs, video sequences, and audio clips.

Elementary Particles
http://www.neutron.anl.gov/Particles.htm

Detailed explanation of elementary particles, 'the most basic physical constituents of the universe', with links to other sources of related information.

Elementistory
http://smallfry.dmu.ac.uk/chem/periodic/elementi.html

Periodic table of elements showing historical rather than scientific information. The contents under the chemical links in the table are mainly brief in nature, providing primarily names and dates of discovery.

Eric's Treasure Trove of Chemistry
http://www.treasure-troves.com/chem/chem.html

Searchable database of facts relating to chemistry. The author has included an A–Z glossary of terms, history, biographies, and concepts of chemistry and chemical engineering.

Essay about Marie and Pierre Curie
http://www.nobel.se/physics/articles/curie/

Extended essay on the life and deeds of Pierre and Marie Curie. The site spans the entirety of the couple's astonishing career and their contribution to the promotion of the understanding of radioactivity. It includes sections on Pierre's and Marie's joint research on radiation phenomena as well as Marie's own work after Pierre's untimely death in 1906: her discovery of the elements radium and polonium, her second Nobel prize, this time in chemistry, her difficulties with the press, and her family life.

Exploring the Material World: Three Classroom Teaching Modules

http://www.lbl.gov/MicroWorlds/module_index.html

Attractive presentation on materials, including an interactive tour of current research in the materials sciences. Choose a module to learn how graphite and diamonds can both be forms of the same element, about Kevlar (the material so strong it can stop bullets), and about the delicate chemical balance of marshes and wetlands.

Fajans on the Concept of Isotopes

http://dbhs.wvusd.k12.ca.us/Chem-History/Fajans-Isotope.html

Extract from Kasimir Fajans paper of 1913 titled *Radioactive Transformations and the Periodic System of The Elements.* The paper describes Fajans's discovery of isotopes and goes on to offer examples of radioactive transformations leading to their production.

Fluorine

http://nobel.scas.bcit.bc.ca/resource/ptable/f.htm

Attractive page describing this highly reactive noble gas, part of a much larger site containing online resources for teachers. Hypertext links show the importance of fluorine in areas such as the development of non-stick coatings and the prevention of tooth decay.

Formaldehyde

http://www.nsc.org/EHC/indoor/formald.htm

Page from the US Environmental Health Center which explains the dangers of formaldehyde, and explains what precautions can be made to decrease exposure to the chemical.

Formulas and Names of Ionic Compounds

http://www.branson.org/depts/science/envrsci/unitchem/Form_and_Name_Ionic.html

Aiming to explain how to write chemical formulas, this site includes a table with the names and charges of common ions, general rules of determining the charges on ions, and test questions on writing formulas for ionic compounds.

Fossil Fuels and their Utilization

http://www.ibis.sfu.ca/~rhlogan/fossil.html

Very detailed information on fossil fuels, including information on the evolutionary stages of coal and the disadvantages of using coal. There are links to a 'Fossil Fuel Centre' and an 'Alternative Energy Menu'.

Fritz Haber: Chemist and Patriot

http://step.sdsc.edu/projects95/chem.in.history/essays/haber.html

Find out about the early 20th-century Prussian chemist whose discovery of a method of synthesizing ammonia led to the development of artificial fertilizers and explosives. Largely historical in content, this biographical page contains a good, short description of the Haber–Bosch process.

Gases

http://www.odu.edu/~chem/genchemhelphomepage/topicreview/bp/ch4/1frame.html

Clear explanation of gases and the states of matter, part of an introductory chemistry site aimed at entry-level university students. Click on the sidebar for information on the properties of gases, gas laws, kinetic molecular theory, and deviation from the ideal gas law.

General Chemistry Chapter Notes

http://web.chem.ufl.edu/~chm2040/Notes/

Intended to supplement a lesson plan from the University of Florida, these notes are a helpful summary of chemistry basics including 'Matter, its properties and measurement', 'Atoms and atomic theory', 'Chemical compounds', and 'Chemical reactions'.

General Chemistry Online

http://marie.frostburg.edu/chem/senese/101/index.shtml

This sophisticated Web site is run by the chemistry department of a US university, but its clear text and comprehensive coverage of general chemistry topics make it worth a visit for keen GCSE-level chemists. Searchable databases of common compounds and important chemical terms are two of the most powerful features of a very comprehensive resource.

Glossary of Terms

http://nobel.scas.bcit.bc.ca/resource/

Common chemistry terms helpfully explained by the British Columbia Institute of Technology in Canada. Part of a much larger site aimed at promoting the use of hypermedia in chemistry teaching, this glossary's clear, ample definitions make it a good revision resource.

G N Lewis and the Covalent Bond
http://dbhs.wvusd.k12.ca.us/Chem-History/Lewis-1916/Lewis-1916.html

Transcript of one of the most important papers in the history of chemistry. In the paper US theoretical chemist G N Lewis forwards his ideas on the shared electron bond, later to become known as the covalent bond. This led to the deeper understanding of the principles of chemical reaction, and also provided science with a mechanism for the production of many compounds.

Introduction to Chemical Equations
http://www.netcomuk.co.uk/~rpeters1/aufceam.htm

Essay that introduces the concept of chemical equations to the reader. The text is accompanied by explanatory diagrams.

Introduction to Organic Chemistry: Frequently Asked Questions
http://antoine.frostburg.edu/chem/senese/101/organic/faq.shtml

Simple answers to 'Frequently Asked Questions' relating to organic chemistry. The page is part of a larger Web site called 'General Chemistry Online!'

Libby, Willard Frank
http://kroeber.anthro.mankato.msus.edu/information/biography/klmno/libby_willard.html

Profile of the Nobel prizewinning US chemist. It traces his academic career and official appointments and the process which led to his discovery of the technique of radiocarbon dating.

Matter and Its Changes
http://library.thinkquest.org/10429/high/matter/matter.htm

Detailed explanation of what matter is and how it undergoes physical and chemical changes. The site covers atoms and molecules; solids, liquids, and gases; and elements, compounds, and mixtures. It is supported by colourful images and diagrams.

Merlin's Academy of Alchemy
http://www.synapses.co.uk/alchemy/index.html

Well-organized distance-learning chemistry course, designed to cover chemistry topics to university level, in language a 13-year-old could understand and avoiding advanced maths. The course, which is not free of charge, comes in the form of a downloadable hypertext book, and e-mail tuition and an exam are available.

Mineral Gallery
http://mineral.galleries.com/

Collection of descriptions and images of minerals, organized by mineral name, class (sulphides, oxides, carbonates, and so on), and grouping (such as gemstones, birth stones, and fluorescent minerals).

Modelling the Periodic Table: An Interactive Simulation
http://www.genesismission.org/educate/scimodule/cosmic/ptable.html

Free software to assist students in studying elements and creating their own model of the periodic table of elements. You can create the simulation online (Shockwave plug-in required) or download it.

Modern Alchemist
http://www.geocities.com/CapeCanaveral/Hangar/9825/index.html

Mine of information for chemistry students, from a site aiming to cover the core chemistry curriculum. Important chemical concepts are explained under categories such as: atoms, molecules, stoichometry, thermodynamics, and nuclear chemistry.

Molecular Expressions: The Pharmaceutical Collection
http://micro.magnet.fsu.edu/pharmaceuticals/index.html

Fascinating collection of images showing what over 100 drugs look like when recrystallized and photographed through a microscope.

Molecular Expressions: The Amino Acid Collection
http://micro.magnet.fsu.edu/aminoacids/index.html

Fascinating collection of images showing what all the known amino acids look like when photographed through a microscope. There is also a detailed article about the different amino acids.

Molecular Expressions: The Pesticide Collection
http://micro.magnet.fsu.edu/pesticides/index.html

Fascinating collection of images showing what pesticides look like when recrystallized and photographed through a microscope. There is also an informative article about pesticides.

Molecule of the Month
http://www.bris.ac.uk/Depts/Chemistry/MOTM/motm.htm

Pages on interesting – and sometimes hypothetical – molecules, contributed by university chemistry departments throughout the world.

Natural Science Pages

http://web.jjay.cuny.edu/~acarpi/NSC/index.htm

Designed for general as well as academic use by John Jay College, City University of New York, this interactive site allows the user to explore concepts in basic science including the scientific method, the nature of matter, and atomic structure.

Online 3D Visualization of Molecular Molecules

http://ecpl.chemistry.uch.gr/~baboukas/Java/3Dmol/General.htm

Large collection of 3D chemical structures to explore. The categories of molecules include 'Natural', 'Inorganics', 'Pharmaceuticals', and 'Biomolecules'.

Organic Chemistry: Introduction

http://www.netcomuk.co.uk/%7Erpeters1/aufloc1.htm

Interesting essay on organic chemistry that includes explanatory diagrams.

Organic Reaction Quizzes and Summaries

http://www.towson.edu/~sweeting/orgrxs/reactsum.htm

At this site you can test your knowledge of basic chemical reactions. There are also hints on how to successfully study chemistry.

Oxidation and Reduction

http://library.thinkquest.org/10784/chem2.html

Introduction to these key electrochemical concepts from the student-created 'ThinkQuest' site. Using tables, equations, and easily understood text, this page explains how oxidation–reduction or redox reactions produce the current that drives batteries. Click forward to access useful pages on other electrochemistry topics such as electrodes and Faraday's constant.

Photosynthesis Directory

http://esg-www.mit.edu/esgbio/ps/psdir.html

Helpful introduction to photosynthesis detailing its 'evolution and discovery' and clear explanations of light and dark reactions. Also includes information and a diagram on the structure and function of the chloroplast.

Physical Properties Of Water And Ice

http://www.nyu.edu/pages/mathmol/modules/water/water_student.html

Well-designed guide to the molecular properties of water in its various states. Click on the introduction for a clear outline of the concepts, then take a look at the various molecular models and select which is water and which is ice. This graphics-rich site is part of a much larger resource for science students.

Plastics
http://www.plasticsusa.com/polylist.html

Datasheets on 49 generic plastic materials, showing their comparative chemical and physical properties.

Polymers and Liquid Crystals
http://plc.cwru.edu/

Online tutorial about two modern physical wonders. The site is divided into a 'Virtual textbook' and a 'Virtual laboratory', with corresponding explanations and experiments.

Polymers: They're Everywhere
http://www.nationalgeographic.com/resources/ngo/education/plastics/index.html

Designed for children, this attractive Web site from the US National Geographic Society is a thought-provoking look at the world of polymers. Using 'clickable' images and lively text, the site moves through the world of natural polymers (amber, silk, and turtle-shell are examples), to consider how we use and abuse human-made polymers, and how we could re-use and recycle them.

Properties and Reactions of Alkanes
http://mychemistrypage.future.easyspace.com/Organic/Alkanes_and_Alkenes/Properties_of_Alkanes.htm

Part of a larger comprehensive site on physical, inorganic, and organic chemistry, this page details the properties of the chemical group alkanes and includes useful graphs and tables.

Reactivity Series of Metals
http://www.faytech.cc.nc.us/~rhorner/chm151c/handouts/reaction/reactive.htm

A simple guide to the reactivity series can be found at this address.

Royal Society of Chemistry
http://www.rsc.org/

Work of the UK society to promote understanding of chemistry and assist its advancement. There are full details of the society's research work, online and print publications, and comprehensive educational programme. All the resources of the largest UK chemistry library can be searched.

Science is Fun in the Lab of Shakhashiri
http://scifun.chem.wisc.edu/

University of Wisconsin-Madison chemistry professor Shakhashiri shares the fun of science through home experiments such as 'Exploring acids and bases with red cabbage'.

Sören Sörenson and pH
http://dbhs.wvusd.k12.ca.us/Chem-History/Sorenson-article.html

Excerpt from a paper on enzymatic processes in which Sörenson defined pH as the relative concentration of hydrogen ions in a solution. This extract from the paper not only explains the origin of the term but also demonstrates its meaning.

States of Matter
http://ull.chemistry.uakron.edu/genobc/Chapter_06/

Summary of the various chemistry concepts associated with the states of matter. Part of a much larger chemistry site from the University of Akron, USA, this page explains kinetic molecular theory and the observed properties of the solid, liquid, and gaseous states. Other features include a set of key equations and an interactive practice exam. Linked page contains advertising.

Table of Elements Drill Game
http://www.edu4kids.com/chem/

How well do you know the periodic table of elements? Test your knowledge here.

Theory of Atoms in Molecules
http://www.chemistry.mcmaster.ca/faculty/bader/aim/

Canadian research paper about atoms that assumes some level of knowledge about the subject. The introductory topics include sections on 'What is an atom?' and 'What is a bond?' and will be of interest to the GCSE student or general reader.

Transition Metals
http://www.odu.edu/~chem/genchemhelphomepage/topicreview/bp/ch12/transframe.html

Concise explanation of transition metals that is part of a larger introduction to common chemistry topics for first-year university students. This well-illustrated page shows the position of the transition metals in the periodic table, and contrasts their properties with those of the main group metals. It contains a table of oxidation states of transition metals, and discusses various bonding theories.

Van't Hoff on Tetrahedral Carbon

http://dbhs.wvusd.k12.ca.us/Chem-History/Van't-Hoff-1874.html

Winner of the first Nobel Prize for Chemistry, van't Hoff here describes the tetrahedral bonding properties of the carbon atom in his paper to the *Archives neerlandaises des sciences exactes et naturelles.* Titled rather lengthily 'A suggestion looking to the extension into space of the structural formulas at present used in chemistry. And a note on the optical activity and the chemical constitution of organic compounds', the paper led to further work on the polarizing properties of carbon and the molecules it forms.

Virtual Experiments

http://neon.chem.ox.ac.uk/vrchemistry/labintro/newdefault.html

Series of interactive chemistry experiments for A-level and university-level students hosted by this site from Oxford University, UK. As well as clear instructions and safety information, the site contains photos of key stages of each experiment. Please note that most of the experiments require specialist equipment and are not suitable for the home. However, the site does contain introductions to various subject areas, such as superconductors and simple inorganic solids.

Water

http://www.bris.ac.uk/Depts/Chemistry/MOTM/water/water.htm

'The aqueous environment influences all of the chemistry that takes place in it', according to this Molecule of the Month page from Bristol University, UK, which takes a challenging look at the chemistry of water. One of the most exciting aspects considered is the discovery of the 'hot life' found in thermal vents on the open floor.

Weathering and Erosion

http://www.hbcumi.cau.edu/tqp/301/301-11/301-11.html

Simple summary of 'Physical or mechanical weathering', 'Chemical weathering', and 'Rates of weathering'.

Web Elements

http://www.webelements.com/

Periodic table on the Web, with 12 different categories of information available for each element – from its physical and chemical characteristics to its electronic configuration.

What is an Element?
http://www.bbc.co.uk/worldservice/sci_tech/features/elementary/index.shtml

Simple information on elements and atomic structure, part of the BBC World Service elementary science Web site. A good starting point for revision, this page defines key concepts such as atomic number.

What is Uranium?
http://www.uic.com.au/uran.htm

Comprehensive and informative page on uranium, its properties and uses, mainly in nuclear reactors and weapons, provided by the Uranium Information Council.

7 Glossary

absolute zero
lowest temperature theoretically possible according to kinetic theory, zero kelvin (0 K), equivalent to –273.15°C/–459.67°F, at which molecules are in their lowest energy state. Although the third law of thermodynamics indicates the impossibility of reaching absolute zero, in practice temperatures of less than a billionth of a degree above absolute zero have been produced. Near absolute zero, the physical properties of some materials change substantially; for example, some metals lose their electrical resistance and become superconducting.

absorption spectroscopy or absorptiometry
technique for determining the identity or amount present of a chemical substance by measuring the amount of electromagnetic radiation the substance absorbs at specific wavelengths; see spectroscopy.

acid
compound that releases hydrogen ions (H^+ or protons) in the presence of an ionizing solvent (usually water). Acids react with bases to form salts, and they act as solvents. Strong acids are corrosive; dilute acids have a sour or sharp taste, although in some organic acids this may be partially masked by other flavour characteristics. The strength of an acid is measured by its hydrogen ion concentration, indicated by the pH value. All acids have a pH below 7.0.

Acids can be classified as monobasic, dibasic, tribasic, and so forth, according to their basicity (the number of hydrogen atoms available to react with a base) and degree of ionization (how many of the available hydrogen atoms dissociate in water). Dilute sulphuric acid is classified as a strong (highly ionized), dibasic acid.

Inorganic acids include boric, carbonic, hydrochloric, hydrofluoric, nitric, phosphoric, and sulphuric. Organic acids include ethanoic (acetic), benzoic, citric, methanoic (formic), lactic, oxalic, and salicylic, as well as complex substances such as nucleic acids and amino acids.

Sulphuric, nitric, and hydrochloric acids are sometimes referred to as the mineral acids. Most naturally occurring acids are found as organic compounds, such as the fatty acids R-COOH and sulphonic acids $R\text{-}SO_3H$, where R is an organic molecular structure.

acid salt
chemical compound formed by the partial neutralization of a dibasic or tribasic acid (one that contains two or three replaceable hydrogen atoms). Although a salt, it still contains replaceable hydrogen, so it may undergo the

typical reactions of an acid. Examples are sodium hydrogen sulphate ($NaHSO_4$) and acid phosphates.

actinide

any of a series of 15 radioactive metallic chemical elements with atomic numbers 89 (actinium) to 103 (lawrencium). Elements 89 to 95 occur in nature; the rest of the series are synthesized elements only. Actinides are grouped together because of their chemical similarities (for example, they are all bivalent), the properties differing only slightly with atomic number. The series is set out in a band in the periodic table of the elements, as are the lanthanides.

actinium (Greek *aktis* 'ray')

white, radioactive, metallic element, the first of the actinide series, symbol Ac, atomic number 89, relative atomic mass 227; it is a weak emitter of high-energy alpha particles.

Actinium occurs with uranium and radium in pitchblende and other ores, and can be synthesized by bombarding radium with neutrons. The longest-lived isotope, Ac-227, has a half-life of 21.8 years (all the other isotopes have very short half-lives). Chemically, it is exclusively trivalent, resembling in its reactions the lanthanides and the other actinides. Actinium was discovered in 1899 by the French chemist André Debierne.

activation analysis

technique used to reveal the presence and amount of minute impurities in a substance or element. A sample of a material that may contain traces of a certain element is irradiated with neutrons, as in a reactor. Gamma rays emitted by the material's radioisotopes have unique energies and relative intensities, similar to the spectral lines from a luminous gas. Measurements and interpretation of the gamma-ray spectrum, using data from standard samples for comparison, provide information on the amounts of impurities present.

activation energy

energy required in order to start a chemical reaction. Some elements and compounds will react together merely by bringing them into contact (spontaneous reaction). For others it is necessary to supply energy in order to start the reaction, even if there is ultimately a net output of energy. This initial energy is the activation energy.

activity series

alternative name for reactivity series.

addition polymerization

polymerization reaction in which a single monomer gives rise to a single polymer, with no other reaction products.

addition reaction

chemical reaction in which the atoms of an element or compound react with a double bond or triple bond in an organic compound by opening up one of the bonds and becoming attached to it, for example:

$$CH_2{=}CH_2 + HCl \rightarrow CH_3CH_2Cl$$

Another example is the addition of hydrogen atoms to unsaturated compounds in vegetable oils to produce margarine. Addition reactions are used to make polymers from alkenes.

adhesive

substance that sticks two surfaces together. Natural adhesives (glues) include gelatin in its crude industrial form (made from bones, hide fragments, and fish offal) and vegetable gums. Synthetic adhesives include thermoplastic and thermosetting resins, which are often stronger than the substances they join; mixtures of epoxy resin and hardener that set by chemical reaction; and elastomeric (stretching) adhesives for flexible joints. Superglues are fast-setting adhesives used in very small quantities.

adipic acid

common name for hexanedioic acid.

affinity

force of attraction (see bond) between atoms that helps to keep them in combination in a molecule. The term is also applied to attraction between molecules, such as those of biochemical significance (for example, between enzymes and substrate molecules). This is the basis for affinity chromatography, by which biologically important compounds are separated.

The atoms of a given element may have a greater affinity for the atoms of one element than for another (for example, hydrogen has a great affinity for chlorine, with which it easily and rapidly combines to form hydrogen chloride, but has little or no affinity for argon).

alcohol

any member of a group of organic chemical compounds characterized by the presence of one or more aliphatic OH (hydroxyl) groups in the molecule, and which form esters with acids. The main uses of alcohols are as solvents for gums, resins, lacquers, and varnishes; in the making of dyes; for essential oils in perfumery; and for medical substances in pharmacy. The alcohol produced naturally in the fermentation process and consumed as part of alcoholic beverages is called ethanol (formerly ethyl alcohol).

Alcohols may be liquids or solids, according to the size and complexity of the molecule. The five simplest alcohols form a series in which the number of carbon and hydrogen atoms increases progressively, each one having an extra CH_2 (methylene) group in the molecule: methanol or wood spirit (methyl alcohol, CH_3OH); ethanol (ethyl alcohol, C_2H_5OH); propanol (propyl alcohol,

C_3H_7OH); butanol (butyl alcohol, C_4H_9OH); and pentanol (amyl alcohol, $C_5H_{11}OH$). The lower alcohols are liquids that mix with water; the higher alcohols, such as pentanol, are oily liquids immiscible with water; and the highest are waxy solids – for example, hexadecanol (cetyl alcohol, $C_{16}H_{33}OH$) and melissyl alcohol ($C_{30}H_{61}OH$), which occur in sperm-whale oil and beeswax, respectively. Alcohols containing the CH_2OH group are primary; those containing CHOH are secondary; while those containing COH are tertiary.

aldehyde
any of a group of organic chemical compounds prepared by oxidation of primary alcohols, so that the OH (hydroxyl) group loses its hydrogen to give an oxygen joined by a double bond to a carbon atom (the aldehyde group, with the formula CHO).

aliphatic compound
any organic chemical compound in which the carbon atoms are joined in straight chains, as in hexane (C_6H_{14}), or in branched chains, as in 2-methylpentane ($CH_3CH(CH_3)CH_2CH_2CH_3$).

Aliphatic compounds have bonding electrons localized within the vicinity of the bonded atoms. Cyclic compounds that do not have delocalized electrons are also aliphatic, as in the alicyclic compound cyclohexane (C_6H_{12}) or the heterocyclic piperidine ($C_5H_{11}N$). Compare aromatic compound.

alkali
base that is soluble in water. Alkalis neutralize acids and are soapy to the touch. The strength of an alkali is measured by its hydrogen ion concentration, indicated by the pH value. They may be divided into strong and weak alkalis: a strong alkali (for example, potassium hydroxide, KOH) ionizes completely when dissolved in water, whereas a weak alkali (for example, ammonium hydroxide, NH_4OH) exists in a partially ionized state in solution. All alkalis have a pH above 7.0.

The hydroxides of metals are alkalis. Those of sodium and potassium are corrosive; both were historically derived from the ashes of plants.

The four main alkalis are sodium hydroxide (caustic soda, NaOH); potassium hydroxide (caustic potash, KOH); calcium hydroxide (slaked lime or limewater, $Ca(OH)_2$); and aqueous ammonia ($NH_{3\,(aq)}$). Their solutions all contain the hydroxide ion OH^-, which gives them a characteristic set of properties.

alkali metal
any of a group of six metallic elements with similar chemical properties: lithium, sodium, potassium, rubidium, caesium, and francium. They form a linked group (Group I) in the periodic table of the elements. They are univalent (have a valency of one) and of very low density (lithium, sodium, and potassium float on water); in general they are reactive, soft, low-melting-point metals. Because of their reactivity they are only found as compounds in nature.

alkaline-earth metal
any of a group of six metallic elements with similar bonding properties: beryllium, magnesium, calcium, strontium, barium, and radium. They form a linked group in the periodic table of the elements. They are strongly basic, bivalent (have a valency of two), and occur in nature only in compounds.

alkaloid
any of a number of physiologically active and frequently poisonous substances contained in some plants. They are usually organic bases and contain nitrogen. They form salts with acids and, when soluble, give alkaline solutions.

Substances in this group are included by custom rather than by scientific rules. Examples include morphine, cocaine, quinine, caffeine, strychnine, nicotine, and atropine.

alkane
member of a group of hydrocarbons having the general formula C_nH_{2n+2}, commonly known as ***paraffins***. As they contain only single covalent bonds, alkanes are said to be saturated. Lighter alkanes, such as methane, ethane, propane, and butane, are colourless gases; heavier ones are liquids or solids. In nature they are found in natural gas and petroleum.

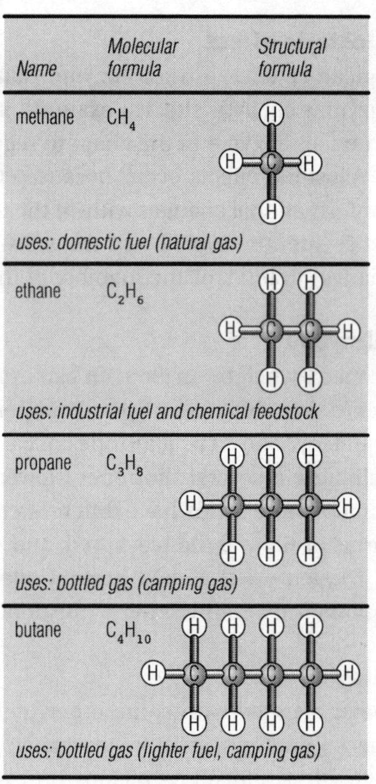

Name
Molecular formula
Structural formula
methane
CH_4
uses: domestic fuel (natural gas)
ethane
C_2H_6
uses: industrial fuel and chemical feedstock
propane
C_3H_8
uses: bottled gas (camping gas)
butane
C_4H_{10}
uses: bottled gas (lighter fuel, camping gas)

The lighter alkanes of methane, ethane, propane, and butane, showing the aliphatic chains, where a hydrogen atom bonds to a carbon atom at all available sites.

alkene
member of the group of hydrocarbons having the general formula C_nH_{2n}, formerly known as ***olefins***. Alkenes are unsaturated com-pounds, characterized by one or more double bonds between adjacent carbon atoms. Lighter alkenes, such as ethene and propene, are gases, obtained from the cracking of oil fractions. Alkenes react by addition, and many useful compounds, such as polyethene (polythene) and bromoethane, are made from them.

alkyl
any organic radical of the formula C_nH_{2n+1}; the chief members are methyl, ethyl, propyl, butyl, and amyl. These radicals are not stable in the free state but are found combined in a large number of types of organic compounds such as alcohols, esters, aldehydes, ketones, and halides.

alkyne

member of the group of hydrocarbons with the general formula C_nH_{2n-2}, formerly known as the ***acetylenes***. They are unsaturated compounds, characterized by one or more triple bonds between adjacent carbon atoms. Lighter alkynes, such as ethyne (formerly acetylene), are gases; heavier ones are liquids or solids.

allene

any of a class of dienes with adjacent double bonds. The simplest example is $CH_2=C=CH_2$, allene itself. Because of the stereochemistry of the double bonds, the terminal hydrogen atoms lie in planes mutually at right angles. Allenes behave mainly as typical unsaturated compounds, but are less stable than dienes with nonadjacent double bonds.

allosteric effect

regulatory effect on an enzyme that takes place at a site distinct from that enzyme's catalytic site. For example, in a chain of enzymes the end product may act on an enzyme in the chain to regulate its own production.

Allosteric effects occur because effector molecules are able to bring about conformational changes with-in the enzyme. This may lead to disruption of the active site, the inability of the substrate molecule (the molecule undergoing change) to bind, or the inability of the products of the reaction to be released.

allotropy

property whereby an element can exist in two or more forms (allo-tropes), each possessing different physical properties but the same state of matter (gas, liquid, or solid). The allotropes of carbon are diamond, fullerene, and graphite. Sulphur has several allotropes (flowers of sulphur, plastic, rhombic, and monoclinic). These solids have different crystal structures, as do the white and grey forms of tin and the black, red, and white forms of phosphorus.

Oxygen exists as two gaseous allotropes: one used by organisms for respiration (O_2), and the other a poisonous pollutant, ozone (O_3).

alloy

metal blended with some other metallic or nonmetallic substance to give it special qualities, such as resistance to corrosion, greater hardness, or tensile strength. Useful alloys include bronze, brass, cupronickel, duralumin, German silver, gunmetal, pewter, solder, steel, and stainless steel.

Among the oldest alloys is bronze (mainly an alloy of copper and tin), the widespread use of which ushered in the Bronze Age. Complex alloys are now common; for example, in dentistry, where a cheaper alternative to gold is made of chromium, cobalt, molybdenum, and titanium. Among the most recent alloys are superplastics: alloys that can stretch to double their length at specific temperatures, permitting, for example, their injection into moulds as easily as plastic.

alpha particle

positively charged, high-energy particle emitted from the nucleus of a radioactive atom. It is one of the products of the spontaneous disintegration of radioactive elements (see radioactivity) such as radium and thorium, and is identical with the nucleus of a helium atom – that is, it consists of two protons and two neutrons. The process of emission, ***alpha decay***, transforms one element into another, decreasing the atomic (or proton) number by two and the atomic mass (or nucleon number) by four.

alum

any double sulphate of a monovalent metal or radical (such as sodium, potassium, or ammonium) and a trivalent metal (such as aluminium, chromium, or iron). The commonest alum is the double sulphate of potassium and aluminium, $K_2Al_2(SO_4)_4.24H_2O$, a white crystalline powder that is readily soluble in water. It is used in curing animal skins. Other alums are used in papermaking and to fix dye in the textile industry.

alumina or corundum

Al_2O_3 oxide of aluminium, widely distributed in clays, slates, and shales. It is formed by the decomposition of the feldspars in granite and used as an abrasive. Typically it is a white powder, soluble in most strong acids or caustic alkalis but not in water. Impure alumina is called 'emery'. Rubies, sapphires, and topaz are corundum gemstones.

aluminium

lightweight, silver-white, ductile and malleable, metallic element, symbol Al, atomic number 13, relative atomic mass 26.9815, melting point 658°C/1,216°F. It is the third most abundant element (and the most abundant metal) in the Earth's crust, of which it makes up about 8.1% by mass. It is non-magnetic, an excellent conductor of electricity, and oxidizes easily; the layer of oxide on its surface making it highly resistant to tarnish.

amalgam

any alloy of mercury with other metals. Most metals will form amalgams, except iron and platinum. Amalgam is used in dentistry for filling teeth, and usually contains copper, silver, and zinc as the main alloying ingredients. This amalgam is pliable when first mixed and then sets hard, but the mercury leaches out and may cause a type of heavy-metal poisoning.

Amalgamation, the process of forming an amalgam, is a technique sometimes used to extract gold and silver from their ores. The ores are ground to a fine sand and brought into contact with mercury, which dissolves the gold and silver particles. The amalgam is then heated to distil off the mercury, leaving a residue of silver and gold. The mercury is recovered and reused.

americium
radioactive metallic element of the actinide series, symbol Am, atomic number 95, relative atomic mass 243.13; it was first synthesized in 1944. It occurs in nature in minute quantities in pitchblende and other uranium ores, where it is produced from the decay of neutron-bombarded plutonium, and is the element with the highest atomic number that occurs in nature. It is synthesized in quantity only in nuclear reactors by the bombardment of plutonium with neutrons. Its longest-lived isotope is Am-243, with a half-life of 7,650 years.

The element was named by Glenn Seaborg, one of the team who first synthesized it in 1944. Ten isotopes are known.

amide
any organic chemical derived from a fatty acid by the replacement of the hydroxyl group (–OH) by an amino group ($–NH_2$). One of the simplest amides is ethanamide (acetamide, CH_3CONH_2), which has a strong mousy odour.

amine
any of a class of organic chemical compounds in which one or more of the hydrogen atoms of ammonia (NH_3) have been replaced by other groups of atoms. ***Methyl amines*** have unpleasant ammonia odours and occur in decomposing fish. They are all gases at ordinary temperature. ***Aromatic amine compounds*** include aniline, which is used in dyeing.

amino acid
water-soluble organic molecule, mainly composed of carbon, oxygen, hydrogen, and nitrogen, containing both a basic amino group (NH_2) and an acidic carboxyl (COOH) group. They are small molecules able to pass through membranes. When two or more amino acids are joined together, they are known as peptides; proteins are made up of peptide chains folded or twisted in characteristic shapes.

ammonia
NH_3 colourless pungent-smelling gas, lighter than air and very soluble in water. It is made on an industrial scale by the Haber (or Haber–Bosch) process, and used mainly to produce nitrogenous fertilizers, nitric acid, and some explosives.

amphoteric
of a chemical compound able to behave either as an acid or as a base depending on its environment. For example, the metals aluminium and zinc, and their oxides and hydroxides, act as bases in acidic solutions and as acids in alkaline solutions. Amino acids and proteins are also amphoteric, as they contain both a basic (amino, $–NH_2$) and an acidic (carboxyl, –COOH) group.

analytical chemistry
branch of chemistry that deals with the determination of the chemical composition of substances. ***Qualitative analysis*** determines the identities of the

substances in a given sample; ***quantitative analysis*** determines how much of a particular substance is present.

Simple qualitative techniques exploit the specific, easily observable properties of elements or compounds – for example, the flame test makes use of the different flame colours produced by metal cations when their compounds are held in a hot flame. More sophisticated methods, such as those of spectroscopy, are required where substances are present in very low concentrations or where several substances have similar properties.

Most quantitative analyses involve initial stages in which the substance to be measured is extracted from the test sample, and purified. The final analytical stages (or 'finishes') may involve measurement of the substance's mass (gravimetry) or volume (volumetry, titrimetry), or a number of techniques initially developed for qualitative analysis, such as fluorescence and absorption spectroscopy, chromatography, electrophoresis, and polarography. Many modern methods enable quantification by means of a detecting device that is integrated into the extraction procedure (as in gas–liquid chromatography).

anhydride
chemical compound obtained by the removal of water from another compound; usually a dehydrated acid. For example, sulphur(VI) oxide (sulphur trioxide, SO_3) is the anhydride of sulphuric acid (H_2SO_4).

anhydrite
naturally occurring anhydrous calcium sulphate ($CaSO_4$). It is used commercially for the manufacture of plaster of Paris and builders' plaster.

anhydrous
of a chemical compound, containing no water. If the water of crystallization is removed from blue crystals of copper(II) sulphate, a white powder (anhydrous copper sulphate) results. Liquids from which all traces of water have been removed are also described as being anhydrous.

aniline (Portuguese *anil* 'indigo')
$C_6H_5NH_2$ or phenylamine one of the simplest aromatic chemicals (a substance related to benzene, with its carbon atoms joined in a ring). When pure, it is a colourless oily liquid; it has a characteristic odour, and turns brown on contact with air. It occurs in coal tar, and is used in the rubber industry and to make drugs and dyes. It is highly poisonous. Aniline was discovered in 1826, and was originally prepared by the dry distillation of indigo, hence its name.

anion
ion carrying a negative charge. During electrolysis, anions in the electrolyte move towards the anode (positive electrode).

An electrolyte, such as the salt zinc chloride ($ZnCl_2$), is dissociated in aqueous solution or in the molten state into doubly charged Zn^{2+} zinc cations and singly-charged Cl^- anions. During electrolysis, the zinc cations flow to the cathode (to become discharged and liberate zinc metal) and the chloride anions flow to the anode (to become discharged and form chlorine gas).

anode

positive electrode of an electrolytic cell, towards which negative particles (anions), usually in solution, are attracted. See electrolysis.

anodizing

process that increases the resistance to corrosion of a metal, such as aluminium, by building up a protective oxide layer on the surface. The natural corrosion resistance of aluminium is provided by a thin film of aluminium oxide; anodizing increases the thickness of this film and thus the corrosion protection.

It is so called because the metal becomes the anode in an electrolytic bath containing a solution of, for example, sulphuric or chromic acid as the electrolyte. During electrolysis oxygen is produced at the anode, where it combines with the metal to form an oxide film.

anthracene

white, glistening, crystalline, tricyclic, aromatic hydrocarbon with a faint blue fluorescence when pure. Its melting point is about 216°C/421°F and its boiling point 351°C/664°F. It occurs in the high-boiling-point fractions of coal tar, where it was discovered in 1832 by the French chemists Auguste Laurent (1808–1853) and Jean Dumas (1800–1884).

antimony

silver-white, brittle, semimetallic element (a metalloid), symbol Sb (from Latin *stibium*), atomic number 51, relative atomic mass 121.75. It occurs chiefly as the sulphide ore stibnite, and is used to make alloys harder; it is also used in photosensitive substances in colour photography, optical electronics, fireproofing, pigments, and medicine. It was employed by the ancient Egyptians in a mixture to protect the eyes from flies.

aqueous solution

solution in which the solvent is water.

argon (Greek *argos* 'idle')

colourless, odourless, nonmetallic, gaseous element, symbol Ar, atomic number 18, relative atomic mass 39.948. It is grouped with the rare gases, and it was long believed not to react with other substances, but observations now indicate that it can be made to combine with boron fluoride to form compounds. It constitutes almost 1% of the Earth's atmosphere, and was discovered in 1894 by British chemists John Rayleigh and William Ramsay after all oxygen and nitrogen had been removed chemically from a sample of air. It is used in electric discharge tubes and argon lasers.

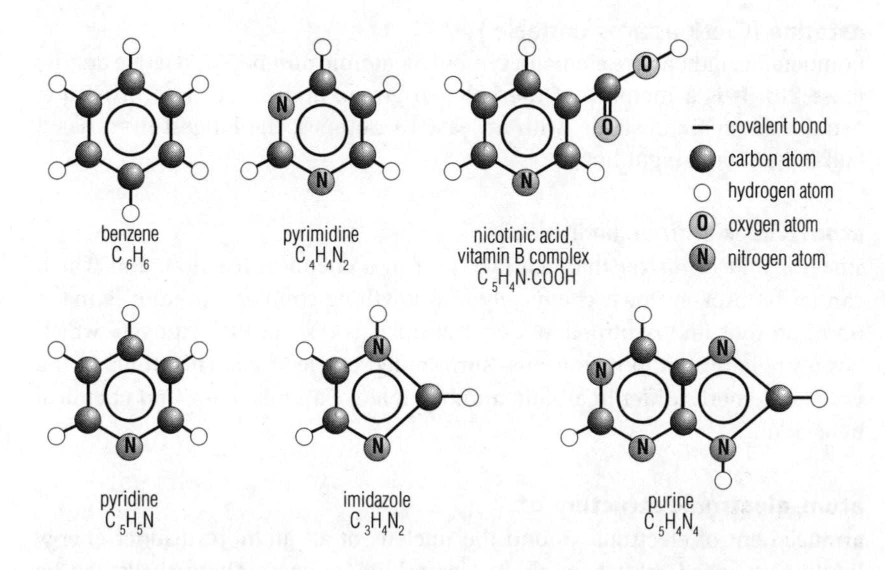

Compounds whose molecules contain the benzene ring, or variations of it, are called aromatic. The term was originally used to distinguish sweet-smelling compounds from others.

aromatic compound

organic chemical compound in which some of the bonding electrons are delocalized (shared among several atoms within the molecule and not localized in the vicinity of the atoms involved in bonding). The commonest aromatic compounds have ring structures, the atoms comprising the ring being either all carbon or mostly carbon with one or more different atoms (usually nitrogen, sulphur, or oxygen). Typical examples are benzene (C_6H_6) and pyridine (C_6H_5N).

arsenic

brittle, greyish-white, semimetallic element (a metalloid), symbol As, atomic number 33, relative atomic mass 74.92. It occurs in many ores and occasionally in its elemental state, and is widely distributed, being present in minute quantities in the soil, the sea, and the human body. In larger quantities, it is poisonous. The chief source of arsenic compounds is as a by-product from metallurgical processes. It is used in making semiconductors, alloys, and solders.

assay

determination of the quantity of a given substance present in a sample. Usually it refers to determining the purity of precious metals. The assay may be carried out by 'wet' methods, when the sample is wholly or partially dissolved in some reagent (often an acid), or by 'dry' or 'fire' methods, in which the compounds present in the sample are combined with other substances.

astatine (Greek *astatos* 'unstable')
nonmetallic, radioactive element, symbol At, atomic number 85, relative atomic mass 210. It is a member of the halogen group, and is very rare in nature. Astatine is highly unstable, with at least 19 isotopes; the longest lived has a half-life of about eight hours.

atom (Greek *atomos* 'undivided')
smallest unit of matter that can take part in a chemical reaction, and which cannot be broken down chemically into anything simpler. An atom is made up of protons and neutrons in a central nucleus (except for hydrogen, which has a single proton in its nucleus) surrounded by electrons. The atoms of the various elements differ in atomic number, relative atomic mass, and chemical behaviour.

atom, electronic structure of
arrangement of electrons around the nucleus of an atom, in distinct energy levels, also called orbitals or shells (see orbital, atomic). These shells can be regarded as a series of concentric spheres, each of which can contain a certain maximum number of electrons; the rare gases have an arrangement in which every shell contains this number (see noble gas structure). The energy levels are usually numbered beginning with the shell nearest to the nucleus. The outermost shell is known as the valency shell as it contains the valence electrons.

The lowest energy level, or innermost shell, can contain no more than two electrons. Outer shells are considered to be stable when they contain eight electrons but additional electrons can sometimes be accommodated provided that the outermost shell has a stable configuration. Electrons in unfilled shells are available to take part in chemical bonding, giving rise to the concept of valency. In ions, the electron shells contain more or fewer electrons than are required for a neutral atom, resulting in negative or positive charges.

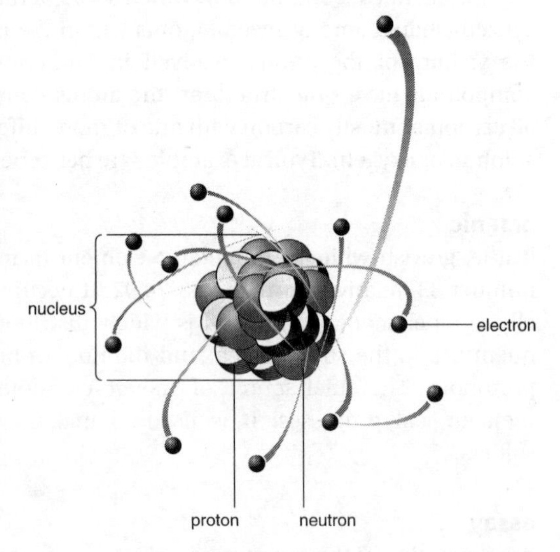

The structure of a sodium atom. The nucleus is composed of 12 protons and 11 neutrons. Eleven electrons orbit the nucleus in 3 orbits: 2 in the inner orbit, 8 in the middle, and 1 in the outer.

The atomic number of an element indicates the number of electrons in a neutral atom. From this it is possible to deduce its electronic structure. For example, sodium has atomic number 11 ($Z = 11$) and its electronic arrangement (configuration) is two electrons in the first energy level, eight electrons in the second energy level and one electron in the third energy level – generally written as 2.8.1. Similarly for sulphur ($Z = 16$), the electron arrangement is 2.8.6. The electronic structure dictates whether two elements will combine by ionic or covalent bonding (see bond) or not at all.

atomicity

number of atoms of an element that combine together to form a molecule. A molecule of oxygen (O_2) has atomicity 2; sulphur (S_8) has atomicity 8.

atomic mass unit or *dalton*; symbol u

unit of mass that is used to measure the relative mass of atoms and molecules. It is equal to one-twelfth of the mass of a carbon-12 atom, which is approximately the mass of a proton or 1.66×10^{-27} kg. The relative atomic mass of an atom has no units; thus oxygen-16 has an atomic mass of 16 daltons but a relative atomic mass of 16.

atomic number or proton number

number (symbol Z) of protons in the nucleus of an atom. It is equal to the positive charge on the nucleus. In a neutral atom, it is also equal to the number of electrons surrounding the nucleus. The chemical elements are arranged in the periodic table of the elements according to their atomic number. See also nuclear notation.

atomic weight

another name for relative atomic mass.

Avogadro's hypothesis

law stating that equal volumes of all gases, when at the same temperature and pressure, have the same numbers of molecules. It was first propounded by Italian chemist Amedeo Avogadro.

Avogadro's number or *Avogadro's constant*

number of carbon atoms in 12 g of the carbon-12 isotope (6.022045×10^{23}). The relative atomic mass of any element, expressed in grams, contains this number of atoms. It is named after Italian chemist Amedeo Avogadro.

azo dye

synthetic dye containing the azo group of two nitrogen atoms (N=N) connecting aromatic ring compounds. Azo dyes are usually red, brown, or yellow, and make up about half the dyes produced. They are manufactured from aromatic amines.

Bakelite
first synthetic plastic, created by Belgian-born US chemist Leo Baekeland in 1909. Bakelite is hard, tough, and heatproof, and is used as an electrical insulator. It is made by the reaction of phenol with methanol (formaldehyde), producing a powdery resin that sets solid when heated. Objects are made by subjecting the resin to compression moulding (simultaneous heat and pressure in a mould).

It is one of the thermosetting plastics, which do not remelt when heated, and is often used for electrical fittings.

barium (Greek *barytes* 'heavy')
soft, silver-white, metallic element, symbol Ba, atomic number 56, relative atomic mass 137.33. It is one of the alkaline-earth metals, found in nature as barium carbonate and barium sulphate. As the sulphate it is used in medicine: taken as a suspension (a 'barium meal'), its movement along the gut is followed using X-rays. The barium sulphate, which is opaque to X-rays, shows the shape of the gut, revealing any abnormalities of the alimentary canal. Barium is also used in alloys, pigments, and safety matches and, with strontium, forms the emissive surface in cathode-ray tubes. It was first discovered in barytes or heavy spar.

base
substance that accepts protons. Bases can contain negative ions such as the hydroxide ion (OH^-), which is the strongest base, or be molecules such as ammonia (NH_3). Ammonia is a weak base, as only some of its molecules accept protons.

$$OH^- + H^+_{(aq)} \rightarrow H_2O_{(l)}$$

$$NH_3 + H_2O \leftrightharpoons NH_4^+ + OH^-$$

Bases that dissolve in water are called alkalis.

Inorganic bases are usually oxides or hydroxides of metals, which react with dilute acids to form a salt and water. Many carbonates also react with dilute acids, additionally giving off carbon dioxide.

basicity
number of replaceable hydrogen atoms in an acid. Nitric acid (HNO_3) is monobasic, sulphuric acid (H_2SO_4) is dibasic, and phosphoric acid (H_3PO_4) is tribasic.

bauxite
principal ore of aluminium, consisting of a mixture of hydrated aluminium oxides and hydroxides, generally contaminated with compounds of iron, which give it a red colour. It is formed by the chemical weathering of rocks in tropical climates. Chief producers of bauxite are Australia, Guinea, Jamaica, Russia, Kazakhstan, Suriname, and Brazil.

To extract aluminium from bauxite, high temperatures (about 800°C/ 1,470°F) are needed to make the ore molten. Strong electric currents are then passed through the molten ore. The process is economical only if cheap electricity is readily available, usually from a hydroelectric plant.

benzaldehyde

C_6H_5CHO colourless liquid with the characteristic odour of almonds. It is used as a solvent and in the making of perfumes and dyes. It occurs in certain leaves, such as the cherry, laurel, and peach, and in a combined form in certain nuts and kernels. It can be extracted from such natural sources, but is usually made from toluene.

benzene

C_6H_6 clear liquid hydrocarbon of characteristic odour, occurring in coal tar. It is used as a solvent and in the synthesis of many chemicals.

The benzene molecule consists of a ring of six carbon atoms, all of which are in a single plane, and it is one of the simplest cyclic compounds. Benzene is the simplest of a class of compounds collectively known as ***aromatic compounds***. Some are considered carcinogenic (cancer-inducing).

benzoic acid

C_6H_5COOH white crystalline solid, sparingly soluble in water, that is used as a preservative for certain foods and as an antiseptic. It is obtained chemically by the direct oxidation of benzaldehyde and occurs in certain natural resins, some essential oils, and as hippuric acid.

berkelium

synthesized, radioactive, metallic element of the actinide series, symbol Bk, atomic number 97, relative atomic mass 247.

It was first produced in 1949 by Glenn Seaborg and his team, at the University of California at Berkeley, USA, after which it is named.

beryllium

hard, light-weight, silver-white, metallic element, symbol Be, atomic number 4, relative atomic mass 9.012. It is one of the alkaline-earth metals, with chemical properties similar to those of magnesium. In nature it is found only in combination with other elements and occurs mainly as beryl ($3BeO.Al_2O_3.6SiO_2$). It is used to make

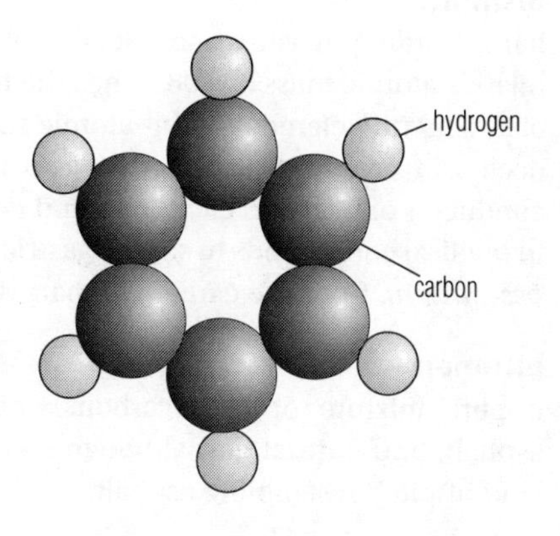

The molecule of benzene consists of six carbon atoms arranged in a ring, with six hydrogen atoms attached. The benzene ring structure is found in many naturally occurring organic compounds.

sturdy, light alloys and to control the speed of neutrons in nuclear reactors. Beryllium oxide was discovered in 1798 by French chemist Louis-Nicolas Vauquelin (1763–1829), but the element was not isolated until 1828, by Friedrich Wöhler and Antoine-Alexandre-Brutus Bussy independently. In 1992 large amounts of beryllium were unexpectedly discovered in six old stars in the Milky Way.

beta particle
electron ejected with great velocity from a radioactive atom that is undergoing spontaneous disintegration. Beta particles do not exist in the nucleus but are created on disintegration, beta decay, when a neutron converts to a proton by emitting an electron.

Beta particles are more penetrating than alpha particles, but less so than gamma radiation; they can travel several metres in air, but are stopped by 2–3 mm of aluminium. They are less strongly ionizing than alpha particles and, like cathode rays, are easily deflected by magnetic and electric fields.

biochemistry
science concerned with the chemistry of living organisms: the structure and reactions of proteins (such as enzymes), nucleic acids, carbohydrates, and lipids. Its study has led to an increased understanding of life processes, such as those by which organisms synthesize essential chemicals from food materials, store and generate energy, and pass on their characteristics through their genetic material. A great deal of medical research is concerned with the ways in which these processes are disrupted. Biochemistry also has applications in agriculture and in the food industry (for instance, in the use of enzymes).

bismuth
hard, brittle, pinkish-white, metallic element, symbol Bi, atomic number 83, relative atomic mass 208.98. It has the highest atomic number of all the stable elements (the elements from atomic number 84 up are radioactive). Bismuth occurs in ores and occasionally as a free metal (native metal). It is a poor conductor of heat and electricity, and is used in alloys of low melting point and in medical compounds to soothe gastric ulcers. The name comes from the Latin *besemutum*, from the earlier German *Wismut*.

bitumen
impure mixture of hydrocarbons, including such deposits as petroleum, asphalt, and natural gas, although sometimes the term is restricted to a soft kind of pitch resembling asphalt.

blast furnace
smelting furnace used to extract metals from their ores, chiefly pig iron from iron ore. The temperature is raised by the injection of an air blast.

In the extraction of iron the ingredients of the furnace are iron ore, coke (carbon), and limestone. The coke is the fuel and provides the carbon monoxide for the reduction of the iron ore; the limestone acts as a flux, removing impurities.

bleaching

decolorization of coloured materials. The two main types of bleaching agent are the ***oxidizing bleaches***, which bring about the oxidation of pigments and include the ultraviolet rays in sunshine, hydrogen peroxide, and chlorine in household bleaches, and the ***reducing bleaches***, which bring about reduction and include sulphur dioxide.

bohrium

synthesized, radioactive element of the transactinide series, symbol Bh, atomic number 107, relative atomic mass 262. It was first synthesized by the Joint Institute for Nuclear Research in Dubna, Russia, in 1976; in 1981 the Laboratory for Heavy Ion Research in Darmstadt, Germany, confirmed its existence. It was named in 1997 after Danish physicist Niels Bohr.

The first chemical study of bohrium was published in 2000, after experiments by Swiss researchers produced six atoms of bohrium-267 (half-life 17 seconds). It behaved like a typical group VII element, with chemical similarities to technetium and rhenium.

boiling point

for any given liquid, the temperature at which the application of heat raises the temperature of the liquid no further, but converts it into vapour.

The boiling point of water under normal pressure is 100°C/212°F. The lower the pressure, the lower the boiling point and vice versa.

bond

result of the forces of attraction that hold together atoms of an element or elements to form a molecule. The principal types of bonding are ionic, covalent, metallic, and intermolecular (such as hydrogen bonding).

The type of bond formed depends on the elements concerned and their electronic structure. In an ionic or electrovalent bond, common in inorganic compounds, the combining atoms gain or lose electrons to become ions; for example, sodium (Na) loses an electron to form a sodium ion (Na^+) while chlorine (Cl) gains an electron to form a chloride ion (Cl^-) in the ionic bond of sodium chloride (NaCl).

In a covalent bond, the atomic orbitals of two atoms overlap to form a molecular orbital containing two electrons, which are thus effectively shared between the two atoms. Covalent bonds are common in organic compounds, such as the four carbon–hydrogen bonds in methane (CH_4). In a dative covalent or coordinate bond, one of the combining atoms supplies both of the valence electrons in the bond.

A metallic bond joins metals in a crystal lattice; the atoms occupy lattice positions as positive ions, and valence electrons are shared between all the ions in an 'electron gas'.

In a hydrogen bond, a hydrogen atom joined to an electronegative atom, such as nitrogen or oxygen, becomes partially positively charged, and is weakly attracted to another electronegative atom on a neighbouring molecule.

boron

nonmetallic element, symbol B, atomic number 5, relative atomic mass 10.811. In nature it is found only in compounds, as with sodium and oxygen in borax. It exists in two allotropic forms (see allotropy): brown amorphous powder and very hard, brilliant crystals. Its compounds are used in the preparation of boric acid, water softeners, soaps, enamels, glass, and pottery glazes. In alloys it is used to harden steel. Because it absorbs slow neutrons, it is used to make boron carbide control rods for nuclear reactors. It is a necessary trace element in the human diet. The element was named by Humphry Davy, who isolated it in 1808, from *bor*ax + -on, as in carb*on*.

brass

metal alloy of copper and zinc, with not more than 5% or 6% of other metals. The zinc content ranges from 20% to 45%, and the colour of brass varies accordingly from coppery to whitish yellow. Brasses are characterized by the ease with which they may be shaped and machined; they are strong and ductile, resist many forms of corrosion, and are used for electrical fittings, ammunition cases, screws, household fittings, and ornaments.

Brasses are usually classed into those that can be worked cold (up to 25% zinc) and those that are better worked hot (about 40% zinc).

bromide

salt of the halide series containing the Br^- ion, which is formed when a bromine atom gains an electron.

The term 'bromide' is sometimes used to describe an organic compound containing a bromine atom, even though it is not ionic. Modern naming uses the term 'bromo-' in such cases. For example, the compound C_2H_5Br is now called bromoethane; its traditional name, still used sometimes, is ethyl bromide.

bromine (Greek *bromos* 'stench')

dark, reddish-brown, nonmetallic element, a volatile liquid at room temperature, symbol Br, atomic number 35, relative atomic mass 79.904. It is a member of the halogen group, has an unpleasant odour, and is very irritating to mucous membranes. Its salts are known as bromides.

Bromine was formerly extracted from salt beds but is now mostly obtained from sea water, where it occurs in small quantities. Its compounds are used in photography and in the chemical and pharmaceutical industries.

buckminsterfullerene

form of carbon, made up of molecules (buckyballs) consisting of 60 carbon atoms arranged in 12 pentagons and 20 hexagons to form a perfect sphere. It was named after the US architect and engineer Richard Buckminster Fuller because of its structural similarity to the geodesic dome that he designed. See fullerene.

buckyballs
popular name for molecules of buckminsterfullerene.

buffer
mixture of compounds chosen to maintain a steady pH. The commonest buffers consist of a mixture of a weak organic acid and one of its salts or a mixture of acid salts of phosphoric acid. The addition of either an acid or a base causes a shift in the chemical equilibrium, thus keeping the pH constant.

butane
C_4H_{10} one of two gaseous alkanes (paraffin hydrocarbons) having the same formula but differing in structure. Normal butane is derived from natural gas; isobutane (2-methylpropane) is a by-product of petroleum manufacture. Liquefied under pressure, it is used as a fuel for industrial and domestic purposes (for example, in portable cookers).

cadmium
soft, silver-white, ductile, and malleable metallic element, symbol Cd, atomic number 48, relative atomic mass 112.40. Cadmium occurs in nature as a sulphide or carbonate in zinc ores. It is a toxic metal that, because of industrial dumping, has become an environmental pollutant. It is used in batteries, electroplating, and as a constituent of alloys used for bearings with low coefficients of friction; it is also a constituent of an alloy with a very low melting point.

Cadmium is also used in the control rods of nuclear reactors, because of its high absorption of neutrons. It was named in 1817 by the German chemist Friedrich Strohmeyer (1776–1835) after the Greek mythological character Cadmus.

caesium (Latin *caesius* 'bluish-grey')
soft, silvery-white, ductile metallic element, symbol Cs, atomic number 55, relative atomic mass 132.905. It is one of the alkali metals, and is the most electropositive of all the elements. In air it ignites spontaneously, and it reacts violently with water. It is used in the manufacture of photocells.

The rate of vibration of caesium atoms has been used as the standard of measuring time. Its radioactive isotope Cs-137 (half-life 30.17 years) is a product of fission in nuclear explosions and in nuclear reactors; it is one of the most dangerous waste products of the nuclear industry, being a highly radioactive biological analogue of potassium. It was named in 1860 by Robert Bunsen, German chemist, from the blueness of its spectral line.

caffeine
alkaloid organic substance found in tea, coffee, and kola nuts; it stimulates the heart and central nervous system. When isolated, it is a bitter crystalline compound, $C_8H_{10}N_4O_2$. Too much caffeine (more than six average cups of tea or coffee a day) can be detrimental to health.

calcination
oxidation of substances by roasting in air.

calcium (Latin *calcis* 'lime')
soft, silvery-white metallic element, symbol Ca, atomic number 20, relative atomic mass 40.08. It is one of the alkaline-earth metals. It is the fifth most abundant element (the third most abundant metal) in the Earth's crust. It is found mainly as its carbonate $CaCO_3$, which occurs in a fairly pure condition as chalk and limestone. Calcium is an essential component of bones, teeth, shells, milk, and leaves, and it forms 1.5% of the human body by mass.

The element was discovered and named by the English chemist Humphry Davy in 1808. Its compounds include slaked lime (calcium hydroxide, $Ca(OH)_2$); plaster of Paris (calcium sulphate, $CaSO_4.2H_2O$); calcium phosphate ($Ca_3(PO_4)_2$), the main constituent of animal bones; calcium hypochlorite ($CaOC1_2$), a bleaching agent; calcium nitrate ($Ca(NO_3)_2.4H_2O$), a nitrogenous fertilizer; calcium carbide (CaC_2), which reacts with water to give ethyne (acetylene); calcium cyanamide ($CaCN_2$), the basis of many pharmaceuticals, fertilizers, and plastics, including melamine; calcium cyanide ($Ca(CN)_2$), used in the extraction of gold and silver and in electroplating; and others used in baking powders and fillers for paints.

calcium carbonate
($CaCO_3$) white solid, found in nature as limestone, marble, and chalk. It is a valuable resource, used in the making of iron, steel, cement, glass, slaked lime, bleaching powder, sodium carbonate and bicarbonate, and many other industrially useful substances.

calcium hydroxide or *slaked lime*
($Ca(OH)_2$) white solid, slightly soluble in water. A solution of calcium hydroxide is called limewater and is used in the laboratory to test for the presence of carbon dioxide. It is manufactured industrially by adding water to calcium oxide (quicklime) in a strongly exothermic reaction.

$$CaO + H_2O \rightarrow Ca(OH)_2$$

It is used to reduce soil acidity and as a cheap alkali in many industrial processes.

californium
synthesized, radioactive, metallic element of the actinide series, symbol Cf, atomic number 98, relative atomic mass 251. It is produced in very small quantities and used in nuclear reactors as a neutron source. The longest-lived isotope, Cf-251, has a half-life of 800 years.

It is named after the state of California, where it was first synthesized in 1950 by US nuclear chemist Glenn Seaborg and his team at the University of California at Berkeley.

camphor

$C_{10}H_{16}O$ volatile, aromatic ketone substance obtained from the camphor tree *Cinnamomum camphora.* It is distilled from chips of the wood, and is used in insect repellents and medicinal inhalants and liniments, and in the manufacture of celluloid.

carbide

compound of carbon and one other chemical element, usually a metal, silicon, or boron. Calcium carbide (CaC_2) can be used as the starting material for many basic organic chemical syntheses, by the addition of water and generation of ethyne (acetylene). Some metallic carbides are used in engineering because of their extreme hardness and strength. Tungsten carbide is an essential ingredient of carbide tools and high-speed tools. The 'carbide process' was used during World War II to make organic chemicals from coal rather than from oil.

carbohydrate

chemical compound composed of carbon, hydrogen, and oxygen, with the basic formula $C_m(H_2O)_n$, and related compounds with the same basic structure but modified functional groups. As sugar and starch, carbohydrates are an important part of a balanced human diet, providing energy for life processes including growth and movement. Excess carbohydrate intake can be converted into fat and stored in the body.

The simplest carbohydrates are sugars (***monosaccharides***, such as glucose and fructose, and ***disaccharides***, such as sucrose), which are soluble compounds, some with a sweet taste. When these basic sugar units are joined together in long chains or branching structures they form ***polysaccharides***, such as starch and glycogen, which often serve as food stores in living organisms. Even more complex carbohydrates are known, including chitin, which is found in the cell walls of fungi and the hard outer skeletons of insects, and cellulose, which makes up the cell walls of plants. Carbohydrates form the chief foodstuffs of herbivorous animals.

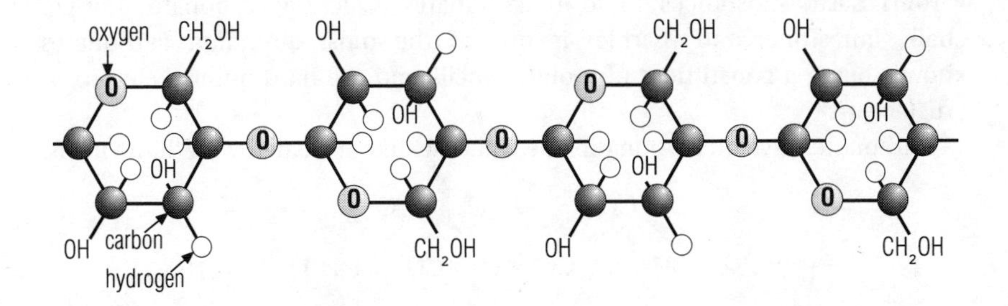

A molecule of the polysaccharide glycogen (animal starch) is formed from linked glucose ($C_6H_{12}O_6$) molecules. A typical glycogen molecule has 100–1,000 glucose units.

carbon (Latin *carbo, carbonaris* 'coal')
nonmetallic element, symbol C, atomic number 6, relative atomic mass 12.011. It occurs on its own as diamond, graphite, and as fullerenes (the allotropes), as compounds in carbonaceous rocks such as chalk and limestone, as carbon dioxide in the atmosphere, as hydrocarbons in petroleum, coal, and natural gas, and as a constituent of all organic substances.

In its amorphous form, it is familiar as coal, charcoal, and soot. The atoms of carbon can link with one another in rings or chains, giving rise to innumerable complex compounds. Of the inorganic carbon compounds, the chief ones are ***carbon dioxide*** (CO_2), a colourless gas formed when carbon is burned in an adequate supply of air; and ***carbon monoxide*** (CO), formed when carbon is oxidized in a limited supply of air. ***Carbon disulphide*** (CS_2) is a dense liquid with a sweetish odour when pure. Another group of compounds is the ***carbon halides***, including carbon tetrachloride (tetrachloromethane, CCl_4).

When added to steel, carbon forms a wide range of alloys with useful properties. In pure form, it is used as a moderator in nuclear reactors; as colloidal graphite it is a good lubricant and, when deposited on a surface in a vacuum, reduces photoelectric and secondary emission of electrons. Carbon is used as a fuel in the form of coal or coke. The radioactive isotope carbon-14 (half-life 5,730 years) is used as a tracer in biological research and in radiocarbon dating. Analysis of interstellar dust has led to the discovery of discrete carbon molecules, each containing 60 carbon atoms. The C_{60} molecules have been named buckminsterfullerenes because of their structural similarity to the geodesic domes designed by US architect and engineer Buckminster Fuller.

carbonate
CO_3^{2-} ion formed when carbon dioxide dissolves in water; any salt formed by this ion and another chemical element, usually a metal.

Carbon dioxide (CO_2) dissolves sparingly in water (for example, when rain falls through the air) to form carbonic acid (H_2CO_3), which unites with various basic substances to form carbonates. Calcium carbonate ($CaCO_3$; chalk, limestone, and marble) is one of the most abundant carbonates known, being a constituent of mollusc shells and the hard outer skeletons of crustaceans.

Carbonates give off carbon dioxide when heated or treated with dilute acids.

$$CO_3^{2-}{}_{(s)} \rightarrow CO_{2\,(g)} + O^{2-}{}_{(s)}$$

$$2H^+{}_{(aq)} + CO_3^{2-}{}_{(s)} \rightarrow CO_{2(g)} + H_2O$$

The latter reaction is used as the laboratory test for the presence of the ion, as it gives an immediate effervescence, with the gas turning limewater (a solution of calcium hydroxide, $Ca(OH)_2$) milky. See sodium carbonate and calcium carbonate.

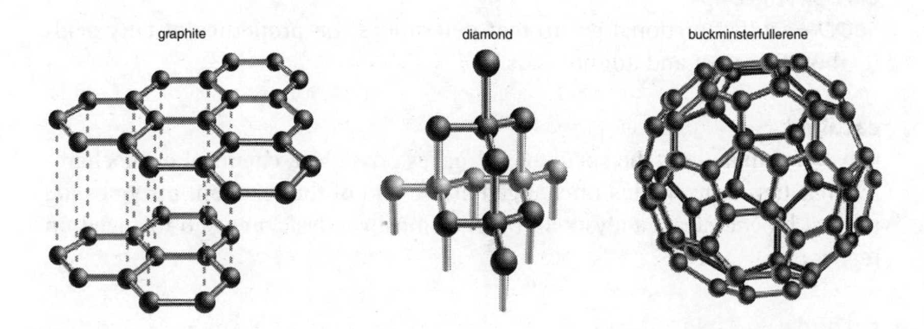

Carbon has three allotropes: diamond, graphite, and the fullerenes. Diamond is strong because each carbon atom is linked to four other carbon atoms. Graphite is made up of layers that slide across one another (giving graphite its qualities as a lubricator); each layer is a giant molecule. In the fullerenes, the carbon atoms form spherical cages. Buckminsterfullerene (shown here) has 60 atoms. Other fullerenes, with 28, 32, 50, 70, and 76 carbon atoms, have also been identified.

carbon dioxide

CO_2 colourless, odourless gas, slightly soluble in water and denser than air. It is formed by the complete oxidation of carbon.

It is produced by living things during the processes of respiration and the decay of organic matter, and plays a vital role in the carbon cycle. It is used as a coolant in its solid form (known as 'dry ice'), and in the chemical industry. Its increasing quantity in the atmosphere contributes to the greenhouse effect and global warming. Britain has 1% of the world's population, yet it produces 3% of CO_2 emissions; the USA has 5% of the world's population and produces 25% of CO_2 emissions.

carbon monoxide

CO colourless, odourless gas formed when carbon is oxidized in a limited supply of air. It is a poisonous constituent of car exhaust fumes, forming a stable compound with haemoglobin in the blood, thus preventing the haemoglobin from transporting oxygen to the body tissues.

In industry, carbon monoxide is used as a reducing agent in metallurgical processes – for example, in the extraction of iron in blast furnaces – and is a constituent of cheap fuels such as water gas. It burns in air with a luminous blue flame to form carbon dioxide.

Carborundum

trademark for a very hard, black abrasive, consisting of silicon carbide (SiC), an artificial compound of carbon and silicon. It is harder than corundum but not as hard as diamond. It was first produced 1891 by US chemist Edward Acheson (1856–1931).

carboxyl group

–COOH acidic functional group that determines the properties of fatty acids (carboxylic acids) and amino acids.

catalyst

substance that alters the speed of, or makes possible, a chemical or biochemical reaction but remains unchanged at the end of the reaction. Enzymes are natural biochemical catalysts. In practice most catalysts are used to speed up reactions.

cathode

negative electrode of an electrolytic cell, towards which positive particles (cations), usually in solution, are attracted. See electrolysis.

A cathode is given its negative charge by connecting it to the negative side of an external electrical supply. This is in contrast to the negative electrode of an electrical (battery) cell, which acquires its charge in the course of a spontaneous chemical reaction taking place within the cell.

cation

ion carrying a positive charge. During electrolysis, cations in the electrolyte move to the cathode (negative electrode).

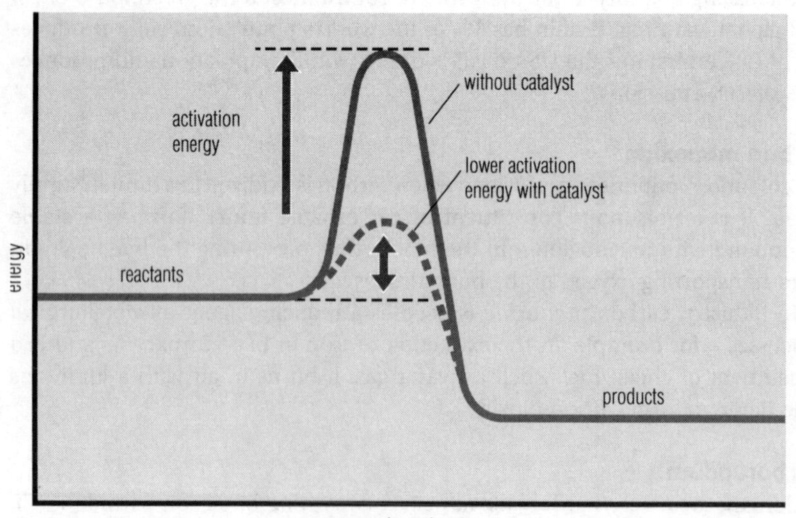

A graph showing how a reaction speeds up with the addition of a catalyst.

cell, electrical or *voltaic cell* or *galvanic cell*
device in which chemical energy is converted into electrical energy; the popular name is 'battery', but this strictly refers to a collection of cells in one unit. The reactive chemicals of a ***primary cell*** cannot be replenished, whereas ***secondary cells*** – such as storage batteries – are rechargeable: their chemical reactions can be reversed and the original condition restored by applying an electric current. It is dangerous to attempt to recharge a primary cell.

cell, electrolytic
device to which electrical energy is applied in order to bring about a chemical reaction; see electrolysis.

celluloid
transparent or translucent, highly flammable, plastic material (a thermoplastic) made from cellulose nitrate and camphor. It was once used for toilet articles, novelties, and photographic film, but has now been replaced by the nonflammable substance cellulose acetate.

cellulose nitrate or *nitrocellulose*
series of esters of cellulose with up to three nitrate (NO_3) groups per monosaccharide unit. It is made by the action of concentrated nitric acid on cellulose (for example, cotton waste) in the presence of concentrated sulphuric acid. Fully nitrated cellulose (gun cotton) is explosive, but esters with fewer nitrate groups were once used in making lacquers, rayon, and plastics, such as coloured and photographic film, until replaced by the nonflammable cellulose acetate. Celluloid is based on cellulose nitrate.

centrifuge
apparatus that rotates containers at high speeds, creating centrifugal forces. One use is for separating mixtures of substances of different densities.

The mixtures are usually spun horizontally in balanced containers ('buckets'), and the rotation sets up centrifugal forces, causing their components to separate according to their densities. A common example is the separation of the lighter plasma from the heavier blood corpuscles in certain blood tests. The ***ultracentrifuge*** is a very high-speed centrifuge, used in biochemistry for separating colloids and organic substances; it may operate at several million revolutions per minute.

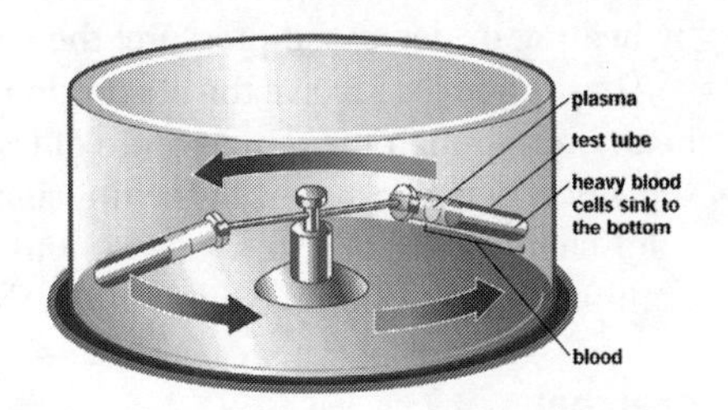

Centrifuge being used to separate the blood plasma from blood cells. As the test tubes spin, the heavier blood cells sink to the bottom. Centrifuges are useful in laboratories and on an industrial scale for separating solids in suspension in liquids.

cerium

malleable and ductile, grey, metallic element, symbol Ce, atomic number 58, relative atomic mass 140.12. It is the most abundant member of the lanthanide series, and is used in alloys, electronic components, nuclear fuels, and lighter flints. It was discovered in 1804 by the Swedish chemists Jöns Berzelius and Wilhelm Hisinger (1766–1852), and, independently, by Martin Klaproth. The element was named after the then recently discovered asteroid Ceres.

chain reaction

succession of reactions, usually involving free radicals, where the products of one stage are the reactants of the next. A chain reaction is characterized by the continual generation of reactive substances.

A chain reaction comprises three separate stages: ***initiation*** – the initial generation of reactive species; ***propagation*** – reactions that involve reactive species and generate similar or different reactive species; and ***termination*** – reactions that involve the reactive species but produce only stable, nonreactive substances. Chain reactions may occur slowly (for example, the oxidation of edible oils) or accelerate as the number of reactive species increases, ultimately resulting in explosion.

chalk

soft, fine-grained, whitish sedimentary rock composed of calcium carbonate, $CaCO_3$, extensively quarried for use in cement, lime, and mortar, and in the manufacture of cosmetics and toothpaste. ***Blackboard chalk*** in fact consists of gypsum (calcium sulphate, $CaSO_4.2H_2O$).

change of state

change in the physical state (solid, liquid, or gas) of a material. For instance, melting, boiling, and evaporation and their opposites solidification and condensation are changes of state. The former set of changes are brought about by heating or decreased pressure; the latter by cooling or increased pressure.

These changes involve the absorption or release of heat energy, called latent heat, even though the temperature of the material does not change during the transition between states. In the unusual change of state called ***sublimation***, a solid changes directly to a gas without passing through the liquid state. For example, solid carbon dioxide (dry ice) sublimes to carbon dioxide gas.

charcoal

black, porous form of carbon, produced by heating wood or other organic materials in the absence of air. It is used as a fuel in the smelting of metals such as copper and zinc, and by artists for making black line drawings. ***Activated charcoal*** has been powdered and dried so that it presents a much increased surface area for adsorption; it is used for filtering and purifying liquids and gases – for example, in drinking-water filters and gas masks.

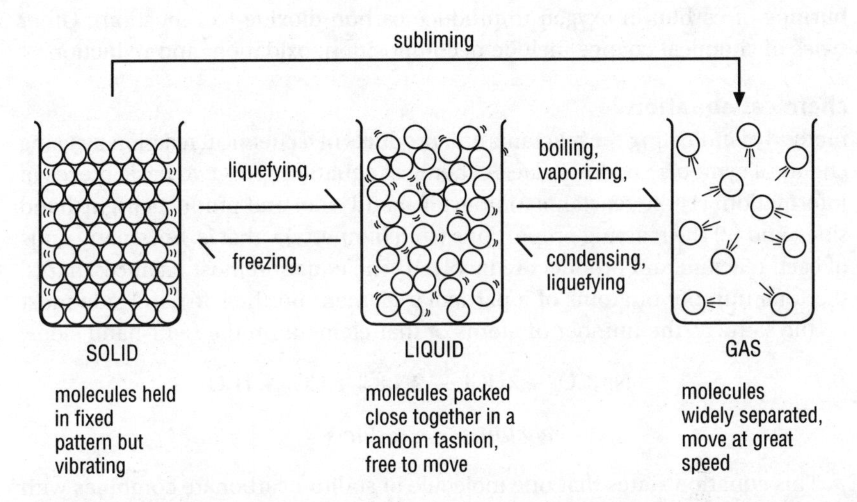

The state (solid, liquid, or gas) of any substance is not fixed but varies with changes in temperature and pressure.

Charles's law

law stating that the volume of a given mass of gas at constant pressure is directly proportional to its absolute temperature (temperature in kelvins). It was discovered by French physicist Jacques Charles in 1787, and independently by French chemist Joseph Gay-Lussac in 1802.

The gas increases by 1/273 (0.003663) of its volume at 0°C for each °C rise of temperature. This means that the coefficient of expansion of all gases is the same. The law is only approximately true.

chelate

chemical compound whose molecules consist of one or more metal atoms or charged ions joined to chains of organic residues by coordinate (or dative covalent) chemical bonds.

The parent organic compound is known as a ***chelating agent*** – for example, EDTA (ethylene-diaminetetra-acetic acid), used in chemical analysis. Chelates are used in analytical chemistry, in agriculture and horticulture as carriers of essential trace metals, in water softening, and in the treatment of thalassaemia by removing excess iron, which may build up to toxic levels in the body. Metalloproteins (natural chelates) may influence the performance of enzymes or provide a mechanism for the storage of iron in the spleen and plasma of the human body.

chemical change

change that occurs when two or more substances (reactants) interact with each other, resulting in the production of different substances (products) with different chemical compositions. A simple example of chemical change is the burning of carbon in oxygen to produce carbon dioxide (combustion). Other types of chemical change include decomposition, oxidation, and reduction.

chemical equation

method of indicating the reactants and products of a chemical reaction by using chemical symbols and formulae. A chemical equation gives two basic pieces of information: (1) the reactants (on the left-hand side) and products (right-hand side); and (2) the reacting proportions (stoichiometry) – that is, how many units of each reactant and product are involved. The equation must balance; that is, the total number of atoms of a particular element on the left-hand side must be the same as the number of atoms of that element on the right-hand side.

$$Na_2CO_3 + 2HCl \rightarrow 2NaCl + CO_2 + H_2O$$

reactants → products

This equation states that one molecule of sodium carbonate combines with two molecules of hydrochloric acid to form two molecules of sodium chloride, one of carbon dioxide, and one of water. Double arrows indicate that the reaction is reversible – in the formation of ammonia from hydrogen and nitrogen, the direction depends on the temperature and pressure of the reactants.

$$3H_2 + N_2 \leftrightharpoons 2NH_3$$

chemical equilibrium

condition in which the products of a reversible chemical reaction (see reversible reaction) are formed at the same rate at which they decompose back into the reactants, so that the concentration of each reactant and product remains constant.

The amounts of reactant and product present at equilibrium are defined by the ***equilibrium constant*** for that reaction and specific temperature.

chemisorption

attachment, by chemical means, of a single layer of molecules, atoms, or ions of gas to the surface of a solid or, less frequently, a liquid. It is the basis of catalysis (see catalyst) and is of great industrial importance.

chemistry

branch of science concerned with the study of the structure and composition of the different kinds of matter, the changes which matter may undergo and the phenomena which occur in the course of these changes.

Organic chemistry is the branch of chemistry that deals with carbon compounds. ***Inorganic chemistry*** deals with the description, properties, reactions, and preparation of all the elements and their compounds, with the exception of carbon compounds. ***Physical chemistry*** is concerned with the quantitative explanation of chemical phenomena and reactions, and the measurement of data required for such explanations. This branch studies in particular the movement of molecules and the effects of temperature and pressure, often with regard to gases and liquids. See also biochemistry.

chloride

Cl^- negative ion formed when hydrogen chloride dissolves in water, and any salt containing this ion, commonly formed by the action of hydrochloric acid (HCl) on various metals or by direct combination of a metal and chlorine. Sodium chloride (NaCl) is common table salt.

chlorinated solvent

any liquid organic compound that contains chlorine atoms, often two or more. These compounds are very effective solvents for fats and greases, but many have toxic properties. They include trichloromethane (chloroform, $CHCl_3$), tetrachloromethane (carbon tetrachloride, CCl_4), and trichloroethene ($CH_2ClCHCl_2$).

chlorination

treatment of water with chlorine in order to disinfect it; also, any chemical reaction in which a chlorine atom is introduced into a chemical compound.

chlorine (Greek *chloros* 'green')

greenish-yellow, gaseous, nonmetallic element with a pungent odour, symbol Cl, atomic number 17, relative atomic mass 35.453. It is a member of the halogen group and is widely distributed, in combination with the alkali metals, as chlorides.

Chlorine was discovered in 1774 by the German chemist Karl Scheele, but English chemist Humphry Davy first proved it to be an element in 1810 and named it after its colour. In nature it is always found in the combined form, as in hydrochloric acid, produced in the mammalian stomach for digestion. Chlorine is obtained commercially by the electrolysis of concentrated brine and is an important bleaching agent and germicide, used for sterilizing both drinking water and swimming pools. As an oxidizing agent it finds many applications in organic chemistry. The pure gas (Cl_2) is a poison and was used in gas warfare in World War I, where its release seared the membranes of the nose, throat, and lungs, producing pneumonia. Chlorine is a component of chlorofluorocarbons (CFCs) and is partially responsible for the depletion of the ozone layer; it is released from the CFC molecule by the action of ultraviolet radiation in the upper atmosphere, making it available to react with and destroy the ozone.

chlorofluorocarbon *CFC*

class of synthetic chemicals that are odourless, nontoxic, nonflammable, and chemically inert. The first CFC was synthesized in 1892, but no use was found for it until the 1920s. Since then their stability and apparently harmless properties have made CFCs popular as propellants in aerosol cans, as refrigerants in refrigerators and air conditioners, as degreasing agents, and in the manufacture of foam packaging. They are partly responsible for the destruction of the ozone layer. In June 1990 representatives of 93 nations, including the UK and the USA, agreed to phase out production of CFCs and various other ozone-depleting chemicals by the end of the 20th century.

When CFCs are released into the atmosphere, they drift up slowly into the stratosphere, where, under the influence of ultraviolet radiation from the Sun, they react with ozone (O_3) to form free chlorine (Cl) atoms and molecular oxygen (O_2), thereby destroying the ozone layer that protects Earth's surface from the Sun's harmful ultraviolet rays. The chlorine liberated during ozone breakdown can react with still more ozone, making the CFCs particularly dangerous to the environment.

chloroform technical name *trichloromethane*

$CHCl_3$ clear, colourless, toxic, carcinogenic liquid with a characteristic pungent, sickly sweet smell and taste, formerly used as an anaesthetic (now superseded by less harmful substances). It is used as a solvent and in the synthesis of organic chemical compounds.

chlorophyll

group of pigments including chlorophyll a and chlorophyll b, the green pigments in plants; it is responsible for the absorption of light energy during photosynthesis. The pigment absorbs the red and blue-violet parts of sunlight but reflects the green, thus giving plants their characteristic colour. Other chlorophylls include chlorophyll c (in brown algae) and chlorophyll d (found in red algae).

Chlorophyll is found within chloroplasts, present in large numbers in leaves. Cyanobacteria (blue-green algae) and other photosynthetic bacteria also have chlorophyll, though of a slightly different type. Chlorophyll is similar in structure to haemoglobin, but with magnesium instead of iron as the reactive part of the molecule.

chromatography (Greek *chromos* 'colour')

technique for separating or analysing a mixture of gases, liquids, or dissolved substances. This is brought about by means of two immiscible substances, one of which (***the mobile phase***) transports the sample mixture through the other (***the stationary phase***). The mobile phase may be a gas or a liquid; the stationary phase may be a liquid or a solid, and may be in a column, on paper, or in a thin layer on a glass or plastic support. The components of the mixture are absorbed or impeded by the stationary phase to different extents and therefore become separated. The technique is used for both qualitative and quantitive analyses in biology and chemistry.

chromium (Greek *chromos* 'colour')

hard, brittle, grey-white, metallic element, symbol Cr, atomic number 24, relative atomic mass 51.996. It takes a high polish, has a high melting point, and is very resistant to corrosion. It is used in chromium electroplating, in the manufacture of stainless steel and other alloys, and as a catalyst. Its compounds are used for tanning leather and for alums. In human nutrition it is a vital trace element. In nature, it occurs chiefly as chrome iron ore or chromite ($FeCr_2O_4$). Kazakhstan, Zimbabwe, and Brazil are sources.

The element was named in 1797 by the French chemist Louis Vauquelin (1763–1829) after its brightly coloured compounds.

citric acid

$HOOCCH_2C(OH)(COOH)CH_2COOH$ organic acid widely distributed in the plant kingdom; it is found in high concentrations in citrus fruits and has a sharp, sour taste. At one time it was commercially prepared from concentrated lemon juice, but now the main source is the fermentation of sugar with certain moulds.

clathrate

compound formed when the small molecules of one substance fill in the holes in the structural lattice of another, solid, substance – for example, sulphur dioxide molecules in ice crystals. Clathrates are therefore intermediate between mixtures and true compounds (which are held together by ionic or covalent chemical bonds).

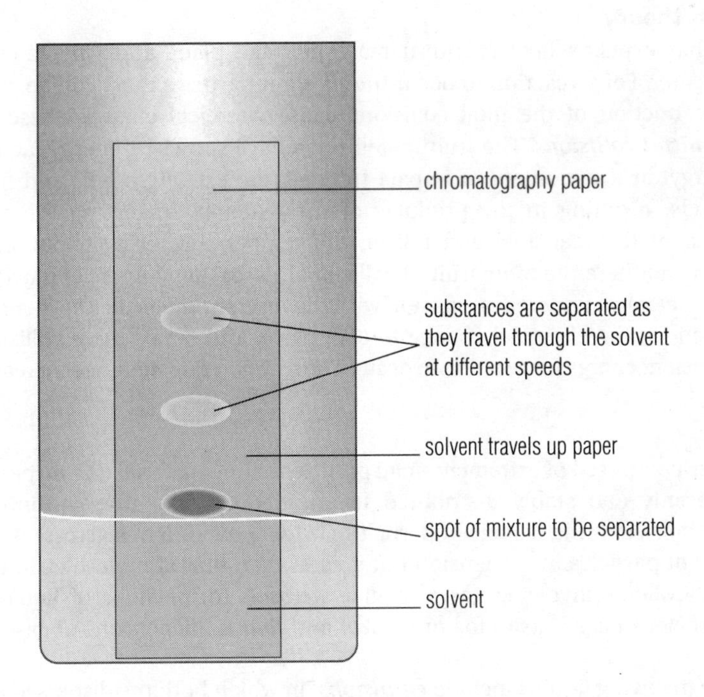

Paper chromatography utilizes the fact that different substances dissolve at different rates. As the solvent travels up the paper it dissolves the mixture, the components of which travel at different speeds and so become separated.

coal tar

black oily material resulting from the destructive distillation of bituminous coal.

Further distillation of coal tar yields a number of fractions: light oil, middle oil, heavy oil, and anthracene oil; the residue is called pitch. On further fractionation a large number of substances are obtained, about 200 of which have been isolated. They are used as dyes and in medicines.

cobalt (German *Kobalt* 'evil spirit')

hard, lustrous, grey, metallic element, symbol Co, atomic number 27, relative atomic mass 58.933. It is found in various ores and occasionally as a free metal, sometimes in metallic meteorite fragments. It is used in the preparation of magnetic, wear-resistant, and high-strength alloys; its compounds are used in inks, paints, and varnishes.

The isotope Co-60 is radioactive (half-life 5.3 years) and is produced in large amounts for use as a source of gamma rays in industrial radiography, research, and cancer therapy. Cobalt was named in 1730 by Swedish chemist Georg Brandt (1694–1768); the name derives from the fact that miners considered its ore malevolent because it interfered with copper production.

collision theory

theory that explains how chemical reactions take place and why rates of reaction alter. For a reaction to occur the reactant particles must collide. Only a certain fraction of the total collisions cause chemical change; these are called ***fruitful collisions***. The fruitful collisions have sufficient energy (activation energy) at the moment of impact to break the existing bonds and form new bonds, resulting in the products of the reaction. Increasing the concentration of the reactants and raising the temperature bring about more collisions and therefore more fruitful collisions, increasing the rate of reaction.

When a catalyst undergoes collision with the reactant molecules, less energy is required for the chemical change to take place, and hence more collisions have sufficient energy for reaction to occur. The reaction rate therefore increases.

colloid

substance composed of extremely small particles of one material (the dispersed phase) evenly and stably distributed in another material (the continuous phase). The size of the dispersed particles (1–1,000 nanometres across) is less than that of particles in suspension but greater than that of molecules in true solution. Colloids involving gases include ***aerosols*** (dispersions of liquid or solid particles in a gas, as in fog or smoke) and ***foams*** (dispersions of gases in liquids).

Those involving liquids include ***emulsions*** (in which both the dispersed and the continuous phases are liquids) and ***sols*** (solid particles dispersed in a liquid). Sols in which both phases contribute to a molecular three-dimensional network have a jellylike form and are known as ***gels***; gelatin, starch 'solution', and silica gel are common examples.

combustion

burning, defined in chemical terms as the rapid combination of a substance with oxygen, accompanied by the evolution of heat and usually light. A slow-burning candle flame and the explosion of a mixture of petrol vapour and air are extreme examples of combustion. Combustion is an exothermic reaction as heat energy is given out.

compound

chemical substance made up of two or more elements bonded together, so that they cannot be separated by physical means. Compounds are held together by ionic or covalent bonds.

The name of a compound may give a clue to its composition. If the name ends in -ide (with the notable exceptions of hydroxides and ammonium chloride) it usually contain two elements. For example calcium oxide is a compound of calcium and oxygen.

If the name ends in -ate or -ite the compound contains oxygen; compounds ending in -ate have a greater proportion of oxygen than those ending in -ite. For example, sodium sulphate (Na_2SO_4) contains more oxygen than does sodium sulphite (Na_2SO_3).

If the name starts with per- the compound contains extra oxygen. For example, hydrogen peroxide H_2O_2 contains one more oxygen than hydrogen oxide (water) H_2O.

The prefix thio- indicates that the compound contains an atom of sulphur in place of an oxygen. For example, sodium thiosulphate ($Na_2S_2O_3$) contains one more sulphur and one less oxygen than the more common sodium sulphate (Na_2SO_4).

The proportions of the different elements in a compound are shown by the chemical formula of that compound. For example, a molecule of sodium sulphate, represented by the formula Na_2SO_4 contains two atoms of sodium, one of sulphur, and four of oxygen.

concentration

amount of a substance (solute) present in a specified amount of a solution. Either amount may be specified as a mass or a volume (liquids only). Common units used are moles per cubic decimetre, grams per cubic decimetre, grams per 100 cubic centimetres, and grams per 100 grams.

The term also refers to the process of increasing the concentration of a solution by removing some of the substance (solvent) in which the solute is dissolved. In a ***concentrated solution***, the solute is present in large quantities. Concentrated brine is around 30% sodium chloride in water; concentrated caustic soda (caustic liquor) is around 40% sodium hydroxide; and concentrated sulphuric acid is 98% acid.

condensation

reaction in which two organic compounds combine to form a larger molecule, accompanied by the removal of a smaller molecule (usually water). This is also known as an addition–elimination reaction. Polyamides (such as nylon) and polyesters (such as Terylene) are made by condensation polymerization.

condensation polymerization

polymerization reaction in which one or more monomers, with more than one reactive functional group, combine to form a polymer with the elimination of water or another small molecule.

condenser

laboratory apparatus used to condense vapours back to liquid so that the liquid can be recovered. It is used in distillation and in reactions where the liquid mixture can be kept boiling without the loss of solvent.

conductor

any material that conducts heat or electricity (as opposed to an insulator, or nonconductor). A good conductor has a high electrical or heat conductivity, and is generally a substance rich in free electrons such as a metal. A poor conductor (such as the nonmetals glass and porcelain) has few free electrons. Carbon is exceptional in being nonmetallic and yet (in some of its forms) a relatively good conductor of heat and electricity. Substances such as silicon and germanium, with intermediate conductivities that are improved by heat, light, or impurities, are known as semiconductors.

conjugation

alternation of double (or triple) and single carbon–carbon bonds in a molecule – for example, in penta-1,3-diene, $H_2C=CH-CH=CH-CH_3$. Conjugation imparts additional stability as the double bonds are less reactive than isolated double bonds.

conservation of energy

principle that states that in a chemical reaction, the total amount of energy in the system remains unchanged. For each component there may be changes in energy due to change of physical state, changes in the nature of chemical bonds, and either an input or output of energy. However, there is no net gain or loss of energy.

constant composition, law of

law that states that the proportions of the amounts of the elements in a pure compound are always the same and are independent of the method by which the compound was produced.

copper

orange-pink, very malleable and ductile, metallic element, symbol Cu (from Latin *cuprum*), atomic number 29, relative atomic mass 63.546. It is used for its durability, pliability, high thermal and electrical conductivity, and resistance to corrosion.

It was the first metal used systematically for tools by humans; when mined and worked into utensils it formed the technological basis for the Copper Age in prehistory. When alloyed with tin it forms bronze, which is stronger than pure copper and may hold a sharp edge; the systematic production and use of this alloy was the basis for the prehistoric Bronze Age. Brass, another hard copper alloy, includes zinc. The element's name comes from the Greek for Cyprus (*Kyprios*), where copper was mined.

corundum

native aluminium oxide, Al_2O_3, the hardest naturally occurring mineral known apart from diamond (corundum rates 9 on the Mohs scale of hardness); lack of cleavage also increases its durability. Its crystals are barrel-shaped prisms of the trigonal system. Varieties of gem-quality corundum are ***ruby*** (red) and ***sapphire*** (any colour other than red, usually blue). Poorer-quality and synthetic corundum is used in industry, for example as an abrasive.

Corundum forms in silica-poor igneous and metamorphic rocks. It is a constituent of emery, which is metamorphosed bauxite.

covalence

form of valence in which two atoms unite by sharing electrons in pairs, so that each atom provides half the shared electrons (see also bond).

covalent bond

chemical bond produced when two atoms share one or more pairs of electrons (usually each atom contributes an electron). The bond is often represented by a single line drawn between the two atoms. Covalently bonded substances include hydrogen (H_2), water (H_2O), and most organic substances.

cracking

reaction in which a large alkane molecule is broken down by heat into a smaller alkane and a small alkene molecule. The reaction is carried out at a high temperature (600°C or higher) and often in the presence of a catalyst. Cracking is a commonly used process in the petrochemical industry.

It is the main method of preparation of alkenes and is also used to manufacture petrol from the higher-boiling-point fractions that are obtained by fractional distillation (fractionation) of crude oil.

crystal

substance with an orderly three-dimensional arrangement of its atoms or molecules, thereby creating an external surface of clearly defined smooth faces having characteristic angles between them. Examples are table salt and quartz.

Each geometrical form, many of which may be combined in one crystal, consists of two or more faces – for example, dome, prism, and pyramid. A mineral can often be identified by the shape of its crystals and the system of crystallization determined. A single crystal can vary in size from a submicroscopic particle to a mass some 30 m/100 ft in length. Crystals fall into seven crystal systems or groups, classified on the basis of the relationship of three or four imaginary axes that intersect at the centre of any perfect, undistorted crystal.

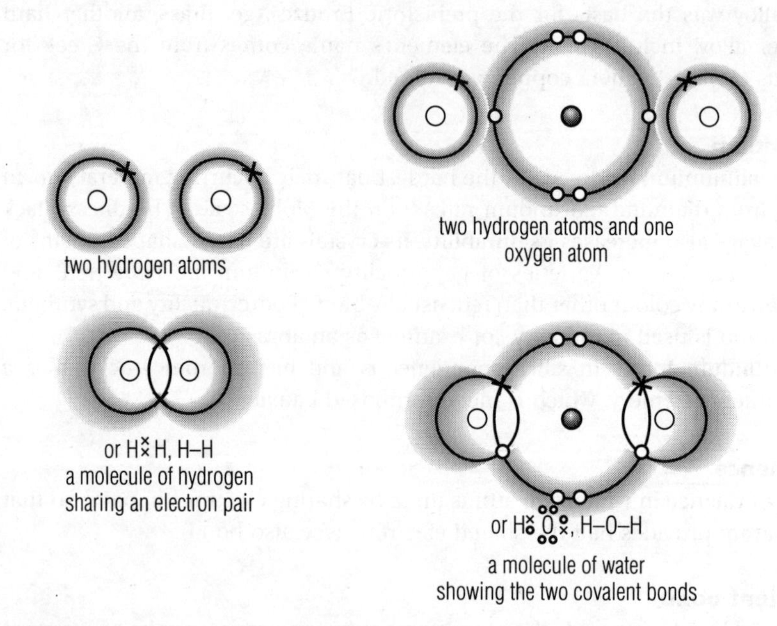

The formation of a covalent bond between two hydrogen atoms to form a hydrogen molecule (H_2), and between two hydrogen atoms and an oxygen atom to form a molecule of water (H_2O). The sharing means that each atom has a more stable arrangement of electrons (its outer electron shells are full).

crystallization
formation of crystals from a liquid, gas, or solution.

crystallography
scientific study of crystals. In 1912 it was found that the shape and size of the repeating atomic patterns (unit cells) in a crystal could be determined by passing X-rays through a sample. This method, known as X-ray diffraction, opened up an entirely new way of 'seeing' atoms. It has been found that many substances have a unit cell that exhibits all the symmetry of the whole crystal; in table salt (sodium chloride, NaCl), for instance, the unit cell is an exact cube.

Many materials were not even suspected of being crystals until they were examined by X-ray crystallography. It has been shown that purified biomolecules, such as proteins and DNA, can form crystals, and such compounds may now be studied by this method. Other applications include the study of metals and their alloys, and of rocks and soils.

cuprite
Cu_2O ore (copper(I) oxide), found in crystalline form or in earthy masses. It is red to black in colour, and is often called ruby copper.

cupronickel
copper alloy (75% copper and 25% nickel), used in hardware products and for coinage.

curium
synthesized, radioactive, metallic element of the *actinide* series, symbol Cm, atomic number 96, relative atomic mass 247. It is produced by bombarding plutonium or americium with neutrons. Its longest-lived isotope has a half-life of 1.7×10^7 years.

cyanide
CN^- ion derived from hydrogen cyanide (HCN), and any salt containing this ion (produced when hydrogen cyanide is neutralized by alkalis), such as potassium cyanide (KCN). The principal cyanides are potassium, sodium, calcium, mercury, gold, and copper. Certain cyanides are poisons. Organic compounds called 'cyanides' (because they contain a CN group) are more properly known as nitrites.

cyclic compound
any of a group of organic chemicals that have rings of atoms in their molecules, giving them a closed-chain structure.

DDT abbreviation for dichloro-diphenyl-trichloroethane
$(ClC_6H_5)_2CHC(HCl_2)$ insecticide discovered in 1939 by Swiss chemist Paul Müller. DDT is highly toxic and persists in the environment and in living tissue. Despite this and its subsequent danger to wildlife, it has evaded a worldwide ban because it remains one of the most effective ways of controlling malaria. China and India were the biggest DDT users in 1999.

decomposition
process whereby a chemical compound is reduced to its component substances. There are three main types of decompositions. Thermal decomposition occurs as a result of heating. For example, copper(II) carbonate decomposes on heating to give copper oxide and carbon dioxide:

$$CuCO_3 \rightarrow CuO + CO_2$$

Electrolytic decomposition may result when an electrical current is passed through a compound in the molten state or in aqueous solution. For example, molten sodium chloride breaks down into sodium and chlorine:

$$2NaCl \rightarrow 2Na + Cl_2$$

Catalysed decomposition describes the process by which decomposition is aided by the presence of a catalyst. For example, hydrogen peroxide decomposes more quickly with the use of manganese(IV) oxide:

$$2H_2O_2 \rightarrow 2H_2O + O_2$$

decrepitation
unusual feature that accompanies the thermal decomposition of some crystals, such as lead(II) nitrate. When these are heated, they spit and crackle and may jump out of the test tube before they decompose.

dehydration
removal of water from a substance to give a product with a new chemical formula; it is not the same as drying.

There are two types of dehydration. For substances such as hydrated copper sulphate ($CuSO_4.5H_2O$) that contain water of crystallization, dehydration means removing this water to leave the anhydrous substance. This may be achieved by heating, and is reversible.

Some substances, such as ethanol, contain the elements of water (hydrogen and oxygen) joined in a different form. ***Dehydrating agents*** such as concentrated sulphuric acid will remove these elements in the ratio 2:1.

deliquescence
phenomenon of a substance absorbing so much moisture from the air that it ultimately dissolves in it to form a solution.

Deliquescent substances make very good drying agents and are used in the bottom chambers of desiccators. Calcium chloride ($CaCl_2$) is one of the commonest.

density
measure of the compactness of a substance; it is equal to its mass per unit volume and is measured in kg per cubic metre/lb per cubic foot. Density is a scalar quantity. The average density D of a mass m occupying a volume V is given by the formula:

$$D = m/V$$

Relative density is the ratio of the density of a substance to that of water at 4°C/39.2°F.

desalination
removal of salt, usually from sea water, to produce fresh water for irrigation or drinking. Distillation has usually been the method adopted, but in the 1970s a cheaper process, using certain polymer materials that filter the molecules of salt from the water by reverse osmosis, was developed.

detergent
surface-active cleansing agent. The common detergents are made from fats (hydrocarbons) and sulphuric acid, and their long-chain molecules have a type of structure similar to that of soap molecules: a salt group at one end attached to a long hydrocarbon 'tail'. They have the advantage over soap in that they do

not produce scum by forming insoluble salts with the calcium and magnesium ions present in hard water.

To remove dirt, which is generally attached to materials by means of oil or grease, the hydrocarbon 'tails' (soluble in oil or grease) penetrate the oil or grease drops, while the 'heads' (soluble in water but insoluble in grease) remain in the water and, being salts, become ionized. Consequently the oil drops become negatively charged and tend to repel one another; thus they remain in suspension and are washed away with the dirt.

deuterium

naturally occurring heavy isotope of hydrogen, mass number 2 (one proton and one neutron), discovered by US chemist Harold Urey in 1932. It is sometimes given the symbol D. In nature, about one in every 6,500 hydrogen atoms is deuterium. Combined with oxygen, it produces 'heavy water' (D_2O), used in the nuclear industry.

deuteron

nucleus of an atom of deuterium (heavy hydrogen). It consists of one proton and one neutron, and is used in the bombardment of chemical elements to synthesize other elements.

diatomic molecule

molecule composed of two atoms joined together. In the case of an element such as oxygen (O_2), the atoms are identical.

dichloro-diphenyl-trichloroethane

full name of the insecticide DDT.

diffusion

spontaneous and random movement of molecules or particles in a fluid (gas or liquid) from a region in which they are at a high concentration to a region of lower concentration, until a uniform concentration is achieved throughout. The difference in concentration between two such regions is called the ***concentration gradient***. No mechanical mixing or stirring is involved. For instance, if a drop of ink is added to water, its molecules will diffuse until their colour becomes evenly distributed throughout. Diffusion occurs more rapidly across a higher concentration gradient and at higher temperature.

dilution

process of reducing the concentration of a solution by the addition of a solvent. The extent of a dilution normally indicates the final volume of solution required. A fivefold dilution would mean the addition of sufficient solvent to make the final volume five times the original.

dioxin

any of a family of over 200 organic chemicals, all of which are heterocyclic hydrocarbons (see cyclic compounds). The term is commonly applied, however, to only one member of the family, 2,3,7,8-tetrachlorodibenzo-*p*-dioxin (2,3,7,8-TCDD), a highly toxic chemical that occurs, for example, as an impurity in the defoliant Agent Orange, used in the Vietnam War, and sometimes in the weed-killer 2,4,5-T. It has been associated with chloracne (a disfiguring skin complaint), birth defects, miscarriages, and cancer.

displacement reaction

chemical reaction in which a less reactive element is replaced in a compound by a more reactive one. For example, the addition of powdered zinc to a solution of copper(II) sulphate displaces copper metal, which can be detected by its characteristic colour:

$$Zn(s) + CuSO_{4(aq)} \rightarrow ZnSO_{4(aq)} + Cu_{(s)}$$

See also electrochemical series.

dissociation

process whereby a single compound splits into two or more smaller products, which may be capable of recombining to form the reactant.

Where dissociation is incomplete (not all the compound's molecules dissociate), a chemical equilibrium exists between the compound and its dissociation products. The extent of incomplete dissociation is defined by a numerical value (dissociation constant).

distillation

technique used to purify liquids or to separate mixtures of liquids possessing different boiling points. ***Simple distillation*** is used in the purification of liquids (or the separation of substances in solution from their solvents) – for example, in the production of pure water from a salt solution.

The solution is boiled and the vapours of the solvent rise into a separate piece of apparatus (the condenser) where they are cooled and condensed. The liquid produced (the distillate) is the pure solvent; the non-volatile solutes (now in solid form) remain in the distillation vessel to be discarded as impurities or recovered as required.

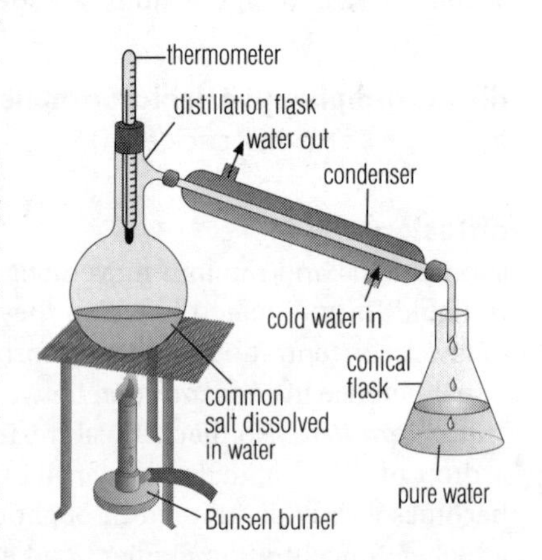

Laboratory apparatus for simple distillation. Other forms of distillation include steam distillation, in which steam is passed into the mixture being distilled, and vacuum distillation, in which air is removed from above the mixture to be distilled.

Mixtures of liquids (such as petroleum or aqueous ethanol) are separated by *fractional distillation*, or fractionation. When the mixture is boiled, the vapours of its most volatile component rise into a vertical fractionating column where they condense to liquid form. However, as this liquid runs back down the column it is reheated to boiling point by the hot rising vapours of the next-most-volatile component and so its vapours ascend the column once more. This boiling-condensing process occurs repeatedly inside the column, eventually bringing about a temperature gradient along its length. The vapours of the more volatile components therefore reach the top of the column and enter the condenser for collection before those of the less volatile components. In the fractional distillation of petroleum, groups of compounds (fractions) possessing similar relative molecular masses and boiling points are tapped off at different points on the column.

double bond

two covalent bonds between adjacent atoms, as in the alkenes (–C=C–) and ketones (–C=O).

double decomposition

reaction between two chemical substances (usually salts in solution) that results in the exchange of a constituent from each compound to create two different compounds. For example, if silver nitrate solution is added to a solution of sodium chloride, there is an exchange of ions yielding sodium nitrate and silver chloride (which is precipitated).

dubnium

synthesized, radioactive, metallic element of the transactinide series, symbol Db, atomic number 105, relative atomic mass 261. Six isotopes have been synthesized, each with very short (fractions of a second) half-lives. Two institutions claim to have been the first to produce it: the Joint Institute for Nuclear Research in Dubna, Russia, in 1967; and the University of California at Berkeley, USA, who disputed the Soviet claim in 1970.

dysprosium (Greek *dusprositos* 'difficult to get near')

silver-white, metallic element of the lanthanide series, symbol Dy, atomic number 66, relative atomic mass 162.50. It is among the most magnetic of all known substances and has a great capacity to absorb neutrons. It was discovered in 1886 by French chemist Paul Lecoq de Boisbaudran (1838–1912).

efflorescence

loss of water or crystallization of crystals exposed to air, resulting in a dry powdery surface.

einsteinium

synthesized, radioactive, metallic element of the actinide series, symbol Es, atomic number 99, relative atomic mass 254.09. It was produced by the first thermonuclear explosion, in 1952, and discovered in fallout debris in the form of the isotope Es-253 (half-life 20 days). Its longest-lived isotope, Es-254, with a half-life of 276 days, allowed the element to be studied at length. It is now synthesized by bombarding lower-numbered transuranic elements in particle accelerators. It was first identified by US chemist Albert Ghiorso and his team who named it in 1955 after Albert Einstein, in honour of his theoretical studies of mass and energy.

elastomer

any material with rubbery properties that stretches easily and then quickly returns to its original length when released. Natural and synthetic rubbers and such materials as polychloroprene and butadiene copolymers are elastomers. The convoluted molecular chains making up these materials are uncoiled by a stretching force, but return to their original position when released because there are relatively few crosslinks between the chains.

electrochemical series or *electromotive series*

list of chemical elements arranged in descending order of the ease with which they can lose electrons to form cations (positive ions). An element can be displaced (displacement reaction) from a compound by any element above it in the series.

electrochemistry

branch of science that studies chemical reactions involving electricity. The use of electricity to produce chemical effects, electrolysis, is employed in many industrial processes, such as electroplating, the manufacture of chlorine, and the extraction of aluminium. The use of chemical reactions to produce electricity is the basis of electrical cells, such as the dry cell and the Leclanché cell.

Since all chemical reactions involve changes to the electronic structure of atoms, all reactions are now recognized as electrochemical in nature. Oxidation, for example, was once defined as a process in which oxygen was combined with a substance, or hydrogen was removed from a compound; it is now defined as a process in which electrons are lost.

Electrochemistry is also the basis of new methods of destroying toxic organic pollutants. For example, the development of electrochemical cells that operate with supercritical water to combust organic waste materials.

electrode

any terminal by which an electric current passes in or out of a conducting substance; for example, the anode or cathode in a battery or the carbons in an arc lamp. The terminals that emit and collect the flow of electrons in thermionic valves (electron tubes) are also called electrodes: for example, cathodes, plates (anodes), and grids.

electrolysis

production of chemical changes by passing an electric current through a solution or molten salt (the electrolyte), resulting in the migration of ions to the electrodes: positive ions (cations) to the negative electrode (cathode) and negative ions (anions) to the positive electrode (anode).

During electrolysis, the ions react with the electrode, either receiving or giving up electrons. The resultant atoms may be liberated as a gas, or deposited as a solid on the electrode, in amounts that are proportional to the amount of current passed, as discovered by English chemist Michael Faraday. For instance, when acidified water is electrolysed, hydrogen ions (H^+) at the cathode receive electrons to form hydrogen gas; hydroxide ions (OH^-) at the anode give up electrons to form oxygen gas and water.

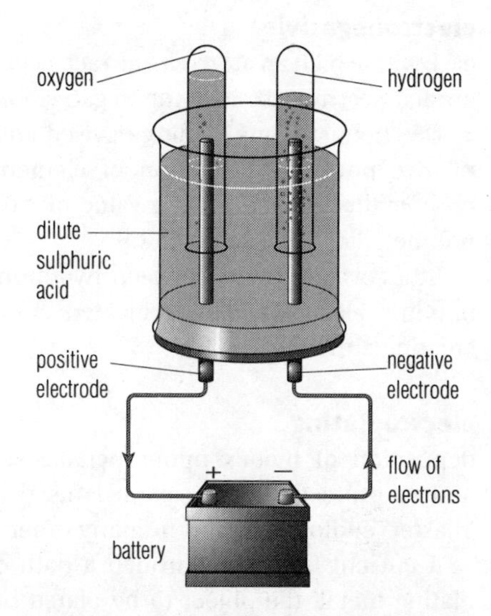

Passing an electric current through acidified water (such as diluted sulphuric acid) breaks down the water into its constituent elements – hydrogen and oxygen.

One application of electrolysis is ***electroplating***, in which a solution of a salt, such as silver nitrate ($AgNO_3$), is the electrolyte and the object to be plated acts as the negative electrode, thus attracting silver ions (Ag^+). Electrolysis is used in many industrial processes, such as coating metals for vehicles and ships, and refining bauxite into aluminium; it also forms the basis of a number of electrochemical analytical techniques, such as polarography.

electrolyte

solution or molten substance in which an electric current is made to flow by the movement and dis-charge of ions in accordance with Faraday's laws of electrolysis.

The term 'electrolyte' is frequently used to denote a substance that, when dissolved in a specified solvent, usually water, dissociates into ions to produce an electrically conducting medium.

electron

stable, negatively charged elementary particle; it is a constituent of all atoms, and a member of the class of particles known as leptons. The electrons in each atom surround the nucleus in groupings called shells; in a neutral atom the number of electrons is equal to the number of protons in the nucleus. This electron structure is responsible for the chemical properties of the atom.

Electrons carry a charge of 1.602177×10^{-19} coulomb and have a mass of 9.109×10^{-31} kg, which is 1/1836 times the mass of a proton. A beam of electrons will undergo diffraction (scattering) and produce interference patterns in the same way as electromagnetic waves such as light; hence they may be regarded as waves as well as particles.

electronegativity

ease with which an atom can attract electrons to itself. Electronegative elements attract electrons, so forming negative ions.

US chemist Linus Pauling devised an electronegativity scale to indicate the relative power of attraction of elements for electrons. Fluorine, the most nonmetallic element, has a value of 4.0 on this scale; oxygen, the next most nonmetallic, has a value of 3.5.

In a covalent bond between two atoms of different electronegativities, the bonding electrons will be located close to the more electronegative atom, creating a dipole.

electroplating

deposition of metals upon metallic surfaces by electrolysis for decorative and/or protective purposes. It is used in the preparation of printing plates, 'master' audio discs, and in many other processes.

A current is passed through a bath containing a solution of a salt of the plating metal, the object to be plated being the cathode (negative terminal); the anode (positive terminal) is either an inert substance or the plating metal. Among the metals most commonly used for plating are zinc, nickel, chromium, cadmium, copper, silver, and gold.

In ***electropolishing***, the object to be polished is made the anode in an electrolytic solution and by carefully controlling conditions the high spots on the surface are dissolved away, leaving a high-quality stain-free surface. This technique is useful in polishing irregular stainless-steel articles.

electropositivity

measure of the ability of elements (mainly metals) to donate electrons to form positive ions. The greater the metallic character, the more electropositive the element.

electrovalent bond

another name for an ionic bond, a chemical bond in which the combining atoms lose or gain electrons to form ions.

element

substance that cannot be split chemically into simpler substances. The atoms of a particular element all have the same number of protons in their nuclei (their atomic, or proton, number). Elements are classified in the periodic table of the elements. Of the known elements, 92 are known to occur in nature on Earth (those with atomic numbers 1–92). Those elements with atomic numbers above 96 do not occur in nature and must be synthesized in particle accelerators. Of the elements, 81 are stable; all the others, which include atomic numbers 43, 61, and from 84 up, are radioactive.

Elements are classified as metals, nonmetals, or metalloids (weakly metallic elements) depending on a combination of their physical and chemical properties; about 75% are metallic. Some elements occur abundantly (oxygen, aluminium); others occur moderately or rarely (chromium, neon); some, in particular the radioactive ones, are found in minute (neptunium, plutonium) or very minute (technetium) amounts. Symbols (devised by Swedish chemist Jöns Berzelius) are used to denote the elements; the symbol is usually the first letter or letters of the English or Latin name (for example, C for carbon, Ca for calcium, Fe for iron, from the Latin *ferrum*). The symbol represents one atom of the element. Two or more elements bonded together form a ***compound*** so that they cannot be separated by physical means. Compounds are held together by ionic or covalent bonds. The number of atoms of an element that combine to form a molecule is it ***atomicity***. A molecule of oxygen (O_2) has atomicity 2; sulphur (S_8) has atomicity 8.

elevation of boiling point

raising of the boiling point of a liquid above that of the pure solvent, caused by a substance being dissolved in it. The phenomenon is observed when salt is added to boiling water; the water ceases to boil because its boiling point has been elevated.

· How much the boiling point is raised depends on the number of molecules of substance dissolved. For a single solvent, such as pure water, all substances in the same molecular concentration produce the same elevation of boiling point. The elevation e produced by the presence of a solute of molar concentration C is given by the equation $e = KC$, where K is a constant (called the ebullioscopic constant) for the solvent concerned.

emergent properties

features of a system that are due to the way in which its components are structured in relation to each other, rather than to the individual properties of those components. Thus the distinctive characteristics of chemical compounds are emergent properties of the way in which the constituent elements are organized, and cannot be explained by the particular properties of those elements taken in isolation.

emission spectroscopy

technique for determining the identity or amount present of a chemical substance by measuring the amount of electromagnetic radiation it emits at specific wavelengths; see spectroscopy.

endothermic reaction

chemical reaction that requires an input of energy in the form of heat for it to proceed; the energy is absorbed from the surroundings by the reactants. The energy absorbed is represented by the symbol + ΔH.

The dissolving of sodium chloride in water and the process of photosynthesis are both endothermic changes. See energy of reaction.

energy of reaction

energy released or absorbed during a chemical reaction, also called ***enthalpy of reaction*** or ***heat of reaction***. In a chemical reaction, the energy stored in the reacting molecules is rarely the same as that stored in the product molecules. Depending on which is the greater, energy is either released (an exothermic reaction) or absorbed (an endothermic reaction) from the surroundings (see conservation of energy). The amount of energy released or absorbed by the quantities of substances represented by the chemical equation is the energy of reaction.

enthalpy

alternative term for energy of reaction, the heat energy associated with a chemical change.

enzyme

biological catalyst produced in cells, and capable of speeding up chemical reactions. They are large, complex proteins, and are highly specific, each chemical reaction requiring its own particular enzyme. The enzyme's specificity arises from its active site, an area with a shape corresponding to part of the molecule with which it reacts (the substrate). The enzyme and the substrate slot together forming an enzyme–substrate complex that allows the reaction to take place, after which the enzyme falls away unaltered.

The activity and efficiency of enzymes are influenced by various factors, including temperature and pH conditions. Temperatures above 60°C/140°F damage (denature) the intricate structure of enzymes, causing reactions to cease. Each enzyme operates best within a specific pH range, and is denatured by excessive acidity or alkalinity.

Enzymes have many medical and industrial uses, from washing powders to drug production, and as research tools in molecular biology. They can be extracted from bacteria and moulds, and genetic engineering now makes it possible to tailor an enzyme for a specific purpose.

equation

representation of a chemical reaction by symbols and numbers; see chemical equation.

erbium

soft, lustrous, greyish, metallic element of the lanthanide series, symbol Er, atomic number 68, relative atomic mass 167.26. It occurs with the element yttrium or as a minute part of various minerals. It was discovered in 1843 by Carl Mosander (1797–1858), and named after the town of Ytterby, Sweden, near which the lanthanides (rare-earth elements) were first found.

Erbium has been used since 1987 to amplify data pulses in optical fibres, enabling faster transmission. Erbium ions in the fibreglass, charged with infrared light, emit energy by amplifying the data pulse as it moves along the fibre.

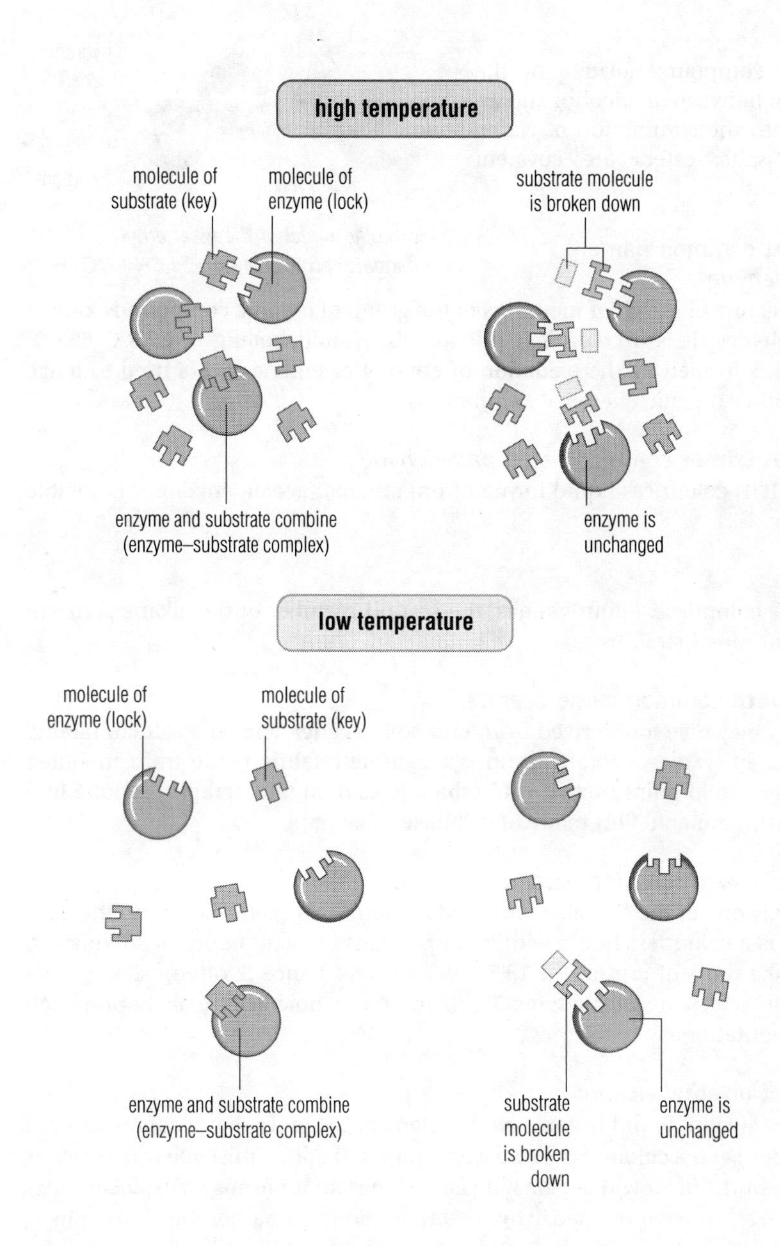

Enzymes are catalysts that can help break larger molecules into smaller molecules while remaining unchanged themselves. Like a key for a lock, each enzyme is specific to one molecule. They also function best within a small temperature and pH range. As the temperature rises enzymes catalyse more molecules but beyond a certain temperature enzymes become denatured.

ester
organic compound formed by the reaction between an alcohol and an acid, with the elimination of water. Unlike salts, esters are covalent compounds.

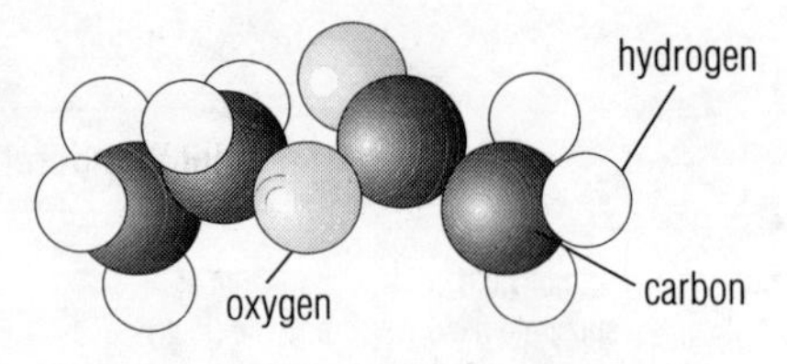

Molecular model of the ester ethyl ethanoate (ethyl acetate) $CH_3CH_2COOCH_3$.

ethanal common name *acetaldehyde*
CH_3CHO one of the chief members of the group of organic compounds known as aldehydes. It is a colourless inflammable liquid boiling at 20.8°C/69.6°F. Ethanal is formed by the oxidation of ethanol or ethene and is used to make many other organic chemical compounds.

ethanal trimer common name *paraldehyde*
$(CH_3CHO)_3$ colourless liquid formed from ethanal (acetaldehyde). It is soluble in water.

ethane
CH_3CH_3 colourless, odourless gas, the second member of the alkane series of hydrocarbons (paraffins).

ethanoate common name *acetate*
$CH_3CO_2^-$ negative ion derived from ethanoic (acetic) acid; any salt containing this ion. In textiles, acetate rayon is a synthetic fabric made from modified cellulose (wood pulp) treated with ethanoic acid; in photography, acetate film is a non-flammable film made of cellulose ethanoate.

ethanoic acid common name *acetic acid*
CH_3CO_2H one of the simplest fatty acids (a series of organic acids). In the pure state it is a colourless liquid with an unpleasant pungent odour; it solidifies to an icelike mass of crystals at 16.7°C/62.4°F, and hence is often called glacial ethanoic acid. Vinegar contains 5% or more ethanoic acid, usually produced by fermentation.

ethanol or *ethyl alcohol*
C_2H_5OH alcohol found in beer, wine, cider, spirits, and other alcoholic drinks. When pure, it is a colourless liquid with a pleasant odour, miscible with water or ether; it burns in air with a pale blue flame. The vapour forms an explosive mixture with air and may be used in high-compression internal combustion engines.

It is produced naturally by the fermentation of carbohydrates by yeast cells. Industrially, it can be made by absorption of ethene and subsequent reaction with water, or by the reduction of ethanal in the presence of a catalyst, and is widely used as a solvent.

Ethanol is used as a raw material in the manufacture of ether, chloral, and iodoform. It can also be added to petrol, where it improves the performance of the engine, or be used as a fuel in its own right (as in Brazil). Crops such as sugar cane may be grown to provide ethanol (by fermentation) for this purpose.

ethene common name *ethylene*

C_2H_4 colourless, flammable gas, the first member of the alkene series of hydrocarbons. It is the most widely used synthetic organic chemical and is used to produce the plastics polyethene (polyethylene), polychloroethene, and polyvinyl chloride (PVC). It is obtained from natural gas or coal gas, or by the dehydration of ethanol.

ether

any of a series of organic chemical compounds having an oxygen atom linking the carbon atoms of two hydrocarbon radical groups (general formula R-O-R′); also the common name for ethoxyethane $C_2H_5OC_2H_5$ (also called diethyl ether). This is used as an anaesthetic and as an external cleansing agent before surgical operations. It is also used as a solvent, and in the extraction of oils, fats, waxes, resins, and alkaloids.

Ethoxyethane is a colourless, volatile, inflammable liquid, slightly soluble in water, and miscible with ethanol. It is prepared by treatment of ethanol with excess concentrated sulphuric acid at 140°C/284°F.

ethyl alcohol

common name for ethanol.

ethylene

common name for ethene.

ethyne common name *acetylene*

CHCH colourless inflammable gas produced by mixing calcium carbide and water. It is the simplest member of the alkyne series of hydrocarbons. It is used in the manufacture of the synthetic rubber neoprene, and in oxyacetylene welding and cutting.

Ethyne was discovered by Edmund Davy in 1836 and was used in early gas lamps, where it was produced by the reaction between water and calcium carbide. Its combustion provides more heat, relatively, than almost any other fuel known (its calorific value is five times that of hydrogen). This means that the gas gives an intensely hot flame; hence its use in oxyacetylene torches.

europium

soft, greyish, metallic element of the lanthanide series, symbol Eu, atomic number 63, relative atomic mass 151.96. It is used in lasers and as the red phosphor in colour televisions; its compounds are used to make control rods for nuclear reactors. It was named in 1901 by French chemist Eugène Demarçay (1852–1904) after the continent of Europe, where it was first found.

exothermic reaction

a chemical reaction during which heat is given out (see energy of reaction). For example, burning sulphur in air to give sulphur dioxide is an exothermic reaction.

$$2S + O_2 \rightarrow SO_2$$

fat

in the broadest sense, a mixture of lipids – chiefly triglycerides (lipids containing three fatty acid molecules linked to a molecule of glycerol). More specifically, the term refers to a lipid mixture that is solid at room temperature (20°C); lipid mixtures that are liquid at room temperature are called ***oils***. The higher the proportion of saturated fatty acids in a mixture, the harder the fat.

fatty acid or *carboxylic acid*

organic compound consisting of a hydrocarbon chain of an even number of carbon atoms, with a carboxyl group (–COOH) at one end. The covalent bonds between the carbon atoms may be single or double; where a double bond occurs the carbon atoms concerned carry one instead of two hydrogen atoms. Chains with only single bonds have all the hydrogen they can carry, so they are said to be saturated with hydrogen. Chains with one or more double bonds are said to be unsaturated. Fatty acids are produced in the small intestine when fat is digested.

Saturated fatty acids include palmitic and stearic acids; unsaturated fatty acids include oleic (one double bond), linoleic (two double bonds), and linolenic (three double bonds). Linoleic acid accounts for more than one third of some margarines. Supermarket brands that say they are high in polyunsaturates may contain as much as 39%. Fatty acids are generally found combined with glycerol in lipids such as triglycerides.

Fehling's test

chemical test to determine whether an organic substance is a reducing agent (substance that donates electrons to other substances in a chemical reaction). It is usually used to detect reducing sugars (monosaccharides, such as glucose, and the disaccharides maltose and lactose) and aldehydes.

If the test substance is heated with a freshly prepared solution containing copper(II) sulphate, sodium hydroxide and sodium potassium tartrate, the production of a brick-red precipitate indicates the presence of a reducing agent.

fermium

synthesized, radioactive, metallic element of the actinide series, symbol Fm, atomic number 100, relative atomic mass 257.10. Ten isotopes are known, the longest-lived of which, Fm-257, has a half-life of 80 days. Fermium has been produced only in minute quantities in particle accelerators. It was discovered in 1952 in the debris of the first thermonuclear explosion. The element was named in 1955 in honour of US physicist Enrico Fermi.

filter

porous substance, such as blotting paper, through which a mixture can be passed to separate out its solid constituents.

filtration

technique by which suspended solid particles in a fluid are removed by passing the mixture through a filter, usually porous paper, plastic, or cloth. The particles are retained by the filter to form a residue and the fluid passes through to make up the filtrate. For example, soot may be filtered from air, and suspended solids from water.

flame test

use of a flame to identify metal cations present in a solid. A nichrome or platinum wire is moistened with acid, dipped in a compound of the element, either powdered or in solution, and then held in a non-luminous hot flame. The colour produced in the flame is characteristic of metals present; for example, sodium burns with an orange-yellow flame and potassium with a lilac one.

flash point

lowest temperature at which a liquid or volatile solid heated under standard conditions gives off sufficient vapour to ignite on the application of a small flame.

The ***fire point*** of a material is the temperature at which full combustion occurs. For safe storage of materials such as fuel or oil, conditions must be well below the flash and fire points to reduce fire risks to a minimum.

fluoride

negative ion (F^-) formed when hydrogen fluoride dissolves in water; compound formed between fluorine and another element in which the fluorine is the more electronegative element (see electronegativity).

fluorine

pale yellow, gaseous, nonmetallic element, symbol F, atomic number 9, relative atomic mass 19. It is the first member of the halogen group of elements, and is pungent, poisonous, and highly reactive, uniting directly with nearly all the elements. It occurs naturally as the minerals fluorite (CaF_2) and cryolite (Na_3AlF_6). Hydrogen fluoride is used in etching glass, and the freons, which all contain fluorine, are widely used as refrigerants.

Fluorine was discovered by the Swedish chemist Karl Scheele in 1771 and isolated by the French chemist Henri Moissan in 1886. Combined with uranium as UF_6, it is used in the separation of uranium isotopes.

fluorocarbon

compound formed by replacing the hydrogen atoms of a hydrocarbon with fluorine. Fluorocarbons are used as inert coatings, refrigerants, synthetic resins, and as propellants in aerosols.

There is concern that the release of fluorocarbons – particularly those containing chlorine (chlorofluorocarbons, CFCs) – depletes the ozone layer, allowing more ultraviolet light from the Sun to penetrate the Earth's atmosphere, and increasing the incidence of skin cancer in humans.

formaldehyde
common name for methanal.

formic acid
common name for methanoic acid.

formula
representation of a molecule, radical, or ion, in which the component chemical elements are represented by their symbols. An ***empirical formula*** indicates the simplest ratio of the elements in a compound, without indicating how many of them there are or how they are combined. A ***molecular formula*** gives the number of each type of element present in one molecule. A ***structural formula*** shows the relative positions of the atoms and the bonds between them. For example, for ethanoic (acetic) acid, the empirical formula is CH_2O, the molecular formula is $C_2H_4O_2$, and the structural formula is CH_3COOH.

fraction
group of similar compounds, the boiling points of which fall within a particular range and which are separated during fractional distillation (fractionation).

fractionation or *fractional distillation*
process used to split complex mixtures (such as petroleum) into their components, usually by repeated heating, boiling, and condensation; see distillation. In the laboratory it is carried out using a fractionating column.

francium
radioactive metallic element, symbol Fr, atomic number 87, relative atomic mass 223. It is one of the alkali metals and occurs in nature in small amounts as a decay product of actinium. Its longest-lived isotope has a half-life of only 21 minutes. Francium was discovered and named in 1939 by Marguérite Perey, to honour her country.

free radical
atom or molecule that has an unpaired electron and is therefore highly reactive. Most free radicals are very short-lived. They are by-products of normal cell chemistry and rapidly oxidize other molecules they encounter. Free radicals are thought to do considerable damage. They are neutralized by protective enzymes.

Free radicals are often produced by high temperatures and are found in flames and explosions.

freezing point, depression of
lowering of a solution's freezing point below that of the pure solvent; it depends on the number of molecules of solute dissolved in it. For a single solvent, such as pure water, all solute substances in the same molar concentration produce

the same lowering of freezing point. The depression *d* produced by the presence of a solute of molar concentration *C* is given by the equation $d = KC$, where *K* is a constant (called the cryoscopic constant) for the solvent concerned. Antifreeze mixtures for car radiators and the use of salt to melt ice on roads are common applications of this principle. Measurement of freezing-point depression is a useful method of determining the molecular weights of solutes. It is also used to detect the illicit addition of water to milk.

fructose

$C_6H_{12}O_6$ sugar that occurs naturally in honey, the nectar of flowers, and many sweet fruits; it is commercially prepared from glucose.

Fructose is a monosaccharide, whereas the more familiar cane or beet sugar is a disaccharide, made up of two monosaccharide units: fructose and glucose. It is sweeter than cane sugar and can be used to sweeten foods for people with diabetes.

fullerene

form of carbon, discovered in 1985, based on closed cages of carbon atoms. The molecules of the most symmetrical of the fullerenes are called buckminsterfullerenes (or buckyballs). They are perfect spheres made up of 60 carbon atoms linked together in 12 pentagons and 20 hexagons, fitted together like those of a spherical football. Other fullerenes with 28, 32, 50, 70, and 76 carbon atoms, have also been identified.

Fullerenes can be made by arcing electricity between carbon rods. They may also occur in candle flames and in clouds of interstellar gas. Fullerene chemistry may turn out to be as important as organic chemistry based on the benzene ring. Already, new molecules based on the buckyball enclosing a metal atom, and 'buckytubes' (cylinders of carbon atoms arranged in hexagons), have been made. They were proved to be 200 times tougher than any other known fibre by Israeli and US materials scientists in 1998. Applications envisaged include using the new molecules as lubricants, semiconductors, and superconductors, and as the starting point for making new drugs.

gadolinium

silvery-white metallic element of the lanthanide series, symbol Gd, atomic number 64, relative atomic mass 157.25. It is found in the products of nuclear fission and used in electronic components, alloys, and products needing to withstand high temperatures.

gallium

grey metallic element, symbol Ga, atomic number 31, relative atomic mass 69.72. It is liquid at room temperature. Gallium arsenide (GaAs) crystals are used for semiconductors in microelectronics, since electrons travel a thousand times faster through them than through silicon. The element was discovered in 1875 by Lecoq de Boisbaudran (1838–1912).

galvanizing
process for rendering iron rust-proof, by plunging it into molten zinc (the dipping method), or by electroplating it with zinc.

gas
form of matter, such as air, in which the molecules move randomly in otherwise empty space, filling any size or shape of container into which the gas is put.

A sugar-lump sized cube of air at room temperature contains 30 trillion molecules moving at an average speed of 500 metres per second (1,800 kph/1,200 mph). Gases can be liquefied by cooling, which lowers the speed of the molecules and enables attractive forces between them to bind them together.

gel
solid produced by the formation of a three-dimensional cage structure, commonly of linked large-molecular-mass polymers, in which a liquid is trapped. It is a form of colloid. A gel may be a jellylike mass (pectin, gelatin) or have a more rigid structure (silica gel).

germanium
brittle, grey-white, weakly metallic (metalloid) element, symbol Ge, atomic number 32, relative atomic mass 72.6. It belongs to the silicon group, and has chemical and physical properties between those of silicon and tin. Germanium is a semiconductor material and is used in the manufacture of transistors and integrated circuits. The oxide is transparent to infrared radiation, and is used in military applications. It was discovered in 1886 by German chemist Clemens Winkler (1838–1904).

glucose or *dextrose* or *grape sugar*
$C_6H_{12}O_6$ sugar present in the blood and manufactured by green plants during photosynthesis. The respiration reactions inside cells involves the oxidation of glucose to produce ATP, the 'energy molecule' used to drive many of the body's biochemical reactions.

glyceride
ester formed between one or more acids and glycerol (propan-1,2,3-triol). A glyceride is termed a mono-, di-, or triglyceride, depending on the number of hydroxyl groups from the glycerol that have reacted with the acids.

Glycerides, chiefly triglycerides, occur naturally as esters of fatty acids in plant oils and animal fats.

glycerine
another name for glycerol.

glycerol or *glycerine* or *propan-1,2,3-triol*

$HOCH_2CH(OH)CH_2OH$ thick, colourless, odourless, sweetish liquid. It is obtained from vegetable and animal oils and fats (by treatment with acid, alkali, superheated steam, or an enzyme), or by fermentation of glucose, and is used in the manufacture of high explosives, in antifreeze solutions, to maintain moist conditions in fruits and tobacco, and in cosmetics.

glycine or *aminoethanoic acid*

$CH_2(NH_2)COOH$ the simplest amino acid, and one of the main components of proteins. When purified, it is a sweet, colourless crystalline compound.

Glycine was found in space in 1994 in the star-forming region Sagittarius B2. The discovery is important because of its bearing on the origins of life on Earth.

glycol or *ethylene glycol* or *ethane-1,2-diol*

$HOCH_2CH_2OH$ thick, colourless, odourless, sweetish liquid. It is used in antifreeze solutions, in the preparation of ethers and esters (used for explosives), as a solvent, and as a substitute for glycerol.

gold

heavy, precious, yellow, metallic element; symbol Au (from Latin *aurum*, 'gold'), atomic number 79, relative atomic mass 197.0. It occurs in nature frequently as a free metal (see native metal) and is highly resistant to acids, tarnishing, and corrosion. Pure gold is the most malleable of all metals and is used as gold leaf or powder, where small amounts cover vast surfaces, such as gilded domes and statues.

The elemental form is so soft that it is alloyed for strength with a number of other metals, such as silver, copper, and platinum. Its purity is then measured in carats on a scale of 24 (24 carats is pure gold). It is used mainly for decorative purposes (jewellery, gilding) but also for coinage, dentistry, and conductivity in electronic devices.

group

vertical column of elements in the periodic table. Elements in a group have similar physical and chemical properties; for example, the group I elements (the alkali metals: lithium, sodium, potassium, rubidium, caesium, and francium) are all highly reactive metals that form univalent ions. There is a gradation of properties down any group: in group I, melting and boiling points decrease, and density and reactivity increase.

Group 0 consists of the rare gases and group II consists of the alkaline-earth metals. Those elements placed between group II and III are the transition metals and group VII contains the halogens.

Haber process or *Haber–Bosch process*

industrial process by which ammonia is manufactured by direct combination of its elements, nitrogen and hydrogen. The reaction is carried out at

400–500°C/752–932°F and at 200 atmospheres pressure. The two gases, in the proportions of 1:3 by volume, are passed over a catalyst of finely divided iron.

Around 10% of the reactants combine, and the unused gases are recycled. The ammonia is separated either by being dissolved in water or by being cooled to liquid form.

$$N_2 + 3H_2 \leftrightharpoons 2NH_3$$

hafnium (Latin *Hafnia* 'Copenhagen')

silvery, metallic element, symbol Hf, atomic number 72, relative atomic mass 178.49. It occurs in nature in ores of zirconium, the properties of which it resembles. Hafnium absorbs neutrons better than most metals, so it is used in the control rods of nuclear reactors; it is also used for light-bulb filaments.

It was named in 1923 by Dutch physicist Dirk Coster (1889–1950) and Hungarian chemist Georg von Hevesy after the city of Copenhagen, where the element was discovered.

half-life

during radioactive decay, the time in which the strength of a radioactive source decays to half its original value. In theory, the decay process is never complete and there is always some residual radioactivity. For this reason, the half-life of a radioactive isotope is measured, rather than the total decay time. It may vary from millionths of a second to billions of years.

Radioactive substances decay exponentially; thus the time taken for the first 50% of the isotope to decay will be the same as the time taken by the next 25%, and by the 12.5% after that, and so on. For example, carbon-14 takes about 5,730 years for half the material to decay; another 5,730 for half of the remaining half to decay; then 5,730 years for half of that remaining half to decay, and so on. Plutonium-239, one of the most toxic of all radioactive substances, has a half-life of about 24,000 years.

halide

any compound produced by the combination of a halogen, such as chlorine or iodine, with a less electronegative element (see electronegativity). Halides may be formed by ionic bonds or by covalent bonds. In organic chemistry, alkyl halides consist of a halogen and alkyl group, such as methyl chloride (chloromethane).

halogen

any of a group of five nonmetallic elements with similar chemical bonding properties: fluorine, chlorine, bromine, iodine, and astatine. They form a linked group in the periodic table of the elements, descending from fluorine, the most reactive, to astatine, the least reactive. They combine directly with most metals to form salts, such as common salt (NaCl). Each halogen has seven electrons in its valence shell, which accounts for the chemical similarities displayed by the group.

halon

organic chemical compound containing one or two carbon atoms, together with bromine and other halogens. The most commonly used are halon 1211 (bromochlorodifluoromethane) and halon 1301 (bromotrifluoromethane). The halons are gases and are widely used in fire extinguishers. As destroyers of the ozone layer, they are up to ten times more effective than chlorofluorocarbons (CFCs), to which they are chemically related.

Levels in the atmosphere are rising by about 25% each year, mainly through the testing of fire-fighting equipment. The use of halons in fire extinguishers was banned in 1994.

hardening of oils

transformation of liquid oils to solid products by hydrogenation. Vegetable oils contain double covalent carbon-to-carbon bonds and are therefore examples of unsaturated compounds. When hydrogen is added to these double bonds, the oils become saturated. The more saturated oils are waxlike solids.

hard water

water that does not lather easily with soap, and produces a deposit or 'scale', in kettles. It is caused by the presence of certain salts of calcium and magnesium.

Temporary hardness is caused by the presence of dissolved hydrogen carbonates (bicarbonates); when the water is boiled, they are converted to insoluble carbonates that precipitate as 'scale'. ***Permanent hardness*** is caused by sulphates and silicates, which are not affected by boiling. Water can be softened by distillation, ion exchange (the principle underlying commercial water softeners), addition of sodium carbonate or of large amounts of soap, or boiling (to remove temporary hardness).

hassium

synthesized, radioactive element of the transactinide series, symbol Hs, atomic number 108, relative atomic mass 265. It was first synthesized in 1984 by the Laboratory for Heavy Ion Research in Darmstadt, Germany.

hazardous waste

waste substance, usually generated by industry, that represents a hazard to the environment or to people living or working nearby. Examples include radioactive wastes, acidic resins, arsenic residues, residual hardening salts, lead from car exhausts, mercury, non-ferrous sludges, organic solvents, asbestos, chlorinated solvents, and pesticides. The cumulative effects of toxic waste can take some time to become apparent (anything from a few hours to many years), and pose a serious threat to the ecological stability of the planet; its economic disposal or recycling is the subject of research.

heat

form of energy possessed by a substance by virtue of the vibrational movement (kinetic energy) of its molecules or atoms. Heat energy is transferred by conduction, convection, and radiation. It always flows from a region of higher temperature (heat intensity) to one of lower temperature. Its effect on a substance may be simply to raise its temperature, or to cause it to expand, melt (if a solid), vaporize (if a liquid), or increase its pressure (if a confined gas).

heat of reaction

alternative term for energy of reaction.

heavy water or *deuterium oxide*

D_2O water containing the isotope deuterium instead of hydrogen (relative molecular mass 20 as opposed to 18 for ordinary water).

Its chemical properties are identical with those of ordinary water, but its physical properties differ slightly. It occurs in ordinary water in the ratio of about one part by mass of deuterium to 5,000 parts by mass of hydrogen, and can be concentrated by electrolysis, the ordinary water being more readily decomposed by this means than the heavy water. It has been used in the nuclear industry because it can slow down fast neutrons, thereby controlling the chain reaction.

helium (Greek *helios* 'Sun')

colourless, odourless, gaseous, non-metallic element, symbol He, atomic number 2, relative atomic mass 4.0026. It is grouped with the rare gases, is nonreactive, and forms no compounds. It is the second most abundant element (after hydrogen) in the universe, and has the lowest boiling (–268.9°C/–452°F) and melting points (–272.2°C/–458°F) of all the elements. It is present in small quantities in the Earth's atmosphere from gases issuing from radioactive elements (from alpha decay) in the Earth's crust; after hydrogen it is the second lightest element. Helium was originally discovered in 1868 in the spectrum of the Sun. It was found on Earth in 1895.

heterogeneous reaction

reaction where there is an interface between the different components or reactants. Examples of heterogeneous reactions are those between a gas and a solid, a gas and a liquid, two immiscible liquids, or two different solids.

hexanedioic acid or *adipic acid*

$(CH_2)_4(COOH)_2$, crystalline solid acid, obtained by the oxidation of certain fatty or waxy bodies. It is a dibasic acid, akin to oxalic and succinic acids, with the typical reactions of carboxylic acids, and is used in the manufacture of nylon 66.

holmium (Latin *Holmia* 'Stockholm')

silvery, metallic element of the lanthanide series, symbol Ho, atomic number 67, relative atomic mass 164.93. It occurs in combination with other rare-earth metals and in various minerals such as gadolinite. Its compounds are highly magnetic.

Alkane	Alcohol	Aldehyde	Ketone	Carboxylic acid	Alkene
CH_4 methane	CH_3OH methanol	$HCHO$ methanal	—	HCO_2H methanoic acid	—
CH_3CH_3 ethane	CH_3CH_2OH ethanol	CH_3CHO ethanal	—	CH_3CO_2H ethanoic acid	CH_2CH_2 ethene
$CH_3CH_2CH_3$ propane	$CH_3CH_2CH_2OH$ propanol	CH_3CH_2CHO propanal	CH_3COCH_3 propanone	$CH_3CH_2CO_2H$ propanoic acid	CH_2CHCH_3 propene
methane	methanol	methanal	propanone	methanoic acid	ethene

Six different types of homologous series: organic chemicals with similar chemical properties in which members differ by a constant relative atomic mass. For example, all the members of the alkane series differ by a relative atomic mass of 14.

The element was discovered in 1878, spectroscopically, by the Swiss chemists J L Soret and Delafontaine, and independently in 1879 by Swedish chemist Per Cleve (1840–1905), who named it after Stockholm, near which it was found.

homologous series

any of a number of series of organic chemicals with similar chemical properties in which members differ by a constant relative molecular mass.

Alkanes (paraffins), alkenes (olefins), and alkynes (acetylenes) form such series in which members differ in mass by 14, 12, and 10 atomic mass units respectively. For example, the alkane homologous series begins with methane (CH_4), ethane (C_2H_6), propane (C_3H_8), butane (C_4H_{10}), and pentane (C_5H_{12}), each member differing from the previous one by a CH_2 group (or 14 atomic mass units).

hydrate

chemical compound that has discrete water molecules combined with it. The water is known as ***water of crystallization*** and the number of water molecules associated with one molecule of the compound is denoted in both its name and chemical formula: for example, $CuSO_4.5H_2O$ is copper(II) sulphate pentahydrate.

hydride

chemical compound containing hydrogen and one other element, and in which the hydrogen is the more electronegative element (see electronegativity).

Hydrides of the more reactive metals may be ionic compounds containing a hydride anion (H^-).

hydrocarbon

any of a class of chemical compounds containing only hydrogen and carbon (for example, the alkanes and alkenes). Hydrocarbons are obtained industrially principally from petroleum and coal tar.

Unsaturated hydrocarbons contain at least one double or triple carbon–carbon bond, whereas saturated hydrocarbons contain only single bonds.

hydrochloric acid

HCl highly corrosive solution of hydrogen chloride (a colourless, acidic gas) in water. The concentrated acid is about 35% hydrogen chloride. The acid is a typical strong, monobasic acid forming only one series of salts, the chlorides. It has many industrial uses, including recovery of zinc from galvanized scrap iron and the production of chlorine. It is also produced in the stomachs of animals for the purposes of digestion.

hydrogenation

addition of hydrogen to an unsaturated organic molecule (one that contains double bonds or triple bonds). It is widely used in the manufacture of margarine and low-fat spreads by the addition of hydrogen to vegetable oils.

Vegetable oils contain double carbon-to-carbon bonds and are therefore examples of unsaturated compounds. When hydrogen is added to these double bonds, the oils become saturated and more solid in consistency.

hydrogen carbonate or *bicarbonate*

compound containing the ion HCO_3^-, an acid salt of carbonic acid (solution of carbon dioxide in water). When heated or treated with dilute acids, it gives off carbon dioxide. The most important compounds are sodium hydrogen carbonate (bicarbonate of soda) and calcium hydrogen carbonate.

hydrolysis

chemical reaction in which the action of water or its ions breaks down a substance into smaller molecules. Hydrolysis occurs in certain inorganic salts in solution, in nearly all non-metallic chlorides, in esters, and in other organic substances. It is one of the mechanisms for the breakdown of food by the body, as in the conversion of starch to glucose.

hydrophilic (Greek 'water-loving')

term describing functional groups with a strong affinity for water, such as the carboxyl group (–COOH).

If a molecule contains both a hydrophilic and a hydrophobic group (a group that repels water), it may have an affinity for both aqueous and nonaqueous molecules. Such compounds are used to stabilize emulsions or as detergents.

hydrophobic (Greek 'water-hating')
term describing functional groups that repel water (the opposite of hydrophilic).

hydroxide
any inorganic chemical compound containing one or more hydroxyl (OH) groups and generally combined with a metal. Hydroxides include sodium hydroxide (caustic soda, NaOH), potassium hydroxide (caustic potash, KOH), and calcium hydroxide (slaked lime, $Ca(OH)_2$).

hydroxyl group
an atom of hydrogen and an atom of oxygen bonded together and covalently bonded to an organic molecule. Common compounds containing hydroxyl groups are alcohols and phenols. In chemical reactions, the hydroxyl group (–OH) frequently behaves as a single entity.

ignition temperature or *fire point*
minimum temperature to which a substance must be heated before it will spontaneously burn independently of the source of heat; for example, ethanol has an ignition temperature of 425°C/798°F and a flash point of 12°C/54°F.

indicator
compound that changes its structure and colour in response to its environment. The commonest chemical indicators detect changes in pH (for example, litmus and universal indicator) or in the oxidation state of a system (redox indicators).

indium (Latin *indicum* 'indigo')
soft, ductile, silver-white, metallic element, symbol In, atomic number 49, relative atomic mass 114.82. It occurs in nature in some zinc ores, is resistant to abrasion, and is used as a coating on metal parts. It was discovered in 1863 by German metallurgists Ferdinand Reich (1799–1882) and Hieronymus Richter (1824–1898), who named it after the two indigo lines of its spectrum.

infrared absorption spectrometry
technique used to determine the mineral or chemical composition of artefacts and organic substances, particularly amber. A sample is bombarded by infrared radiation, which causes the atoms in it to vibrate at frequencies characteristic of the substance present, and absorb energy at those frequencies from the infrared spectrum, thus forming the basis for identification.

inorganic chemistry
branch of chemistry dealing with the chemical properties of the elements and their compounds, excluding the more complex covalent compounds of carbon, which are considered in organic chemistry.

The origins of inorganic chemistry lay in observing the characteristics and experimenting with the uses of the substances (compounds and elements) that could be extracted from mineral ores. These could be classified according to their chemical properties: elements could be classified as metals or nonmetals; compounds as acids or bases, oxidizing or reducing agents, ionic compounds (such as salts), or covalent compounds (such as gases). The arrangement of elements into groups possessing similar properties led to Mendeleyev's periodic table of the elements, which prompted chemists to predict the properties of undiscovered elements that might occupy gaps in the table. This, in turn, led to the discovery of new elements, including a number of highly radioactive elements that do not occur naturally.

inorganic compound
compound that does not contain carbon and is not manufactured by living organisms. Water, sodium chloride, and potassium are inorganic compounds because they are widely found outside living cells. However, carbon dioxide is considered inorganic, contains carbon, and is manufactured by organisms during respiration. Carbonates and carbon monoxide are also regarded as inorganic compounds. See organic compound.

intermolecular force or *van der Waals' force*
force of attraction between molecules. Intermolecular forces are relatively weak; hence simple molecular compounds are gases, liquids, or low-melting-point solids.

iodine (Greek *iodes* 'violet')
greyish-black nonmetallic element, symbol I, atomic number 53, relative atomic mass 126.9044. It is a member of the halogen group. Its crystals give off, when heated, a violet vapour with an irritating odour resembling that of chlorine. It occurs only in combination with other elements. Its salts are known as iodides and are found in sea water. As a mineral nutrient it is vital to the proper functioning of the thyroid gland, where it occurs in trace amounts as part of the hormone thyroxine. Absence of iodine from the diet leads to goitre. Iodine is used in photography, in medicine as an antiseptic, and in making dyes.

Its radioactive isotope ^{131}I (half-life of eight days) is a dangerous fission product from nuclear explosions and from the nuclear reactors in power plants, since, if ingested, it can be taken up by the thyroid and damage it. It was discovered in 1811 by French chemist B Courtois (1777–1838).

ion
atom, or group of atoms, that is either positively charged (cation) or negatively charged (anion), as a result of the loss or gain of electrons during chemical reactions or exposure to certain forms of radiation. In solution or in the molten state, ionic compounds such as salts, acids, alkalis, and metal oxides conduct electricity. These compounds are known as electrolytes.

Ions are produced during electrolysis, for example the salt zinc chloride ($ZnCl_2$) dissociates into the positively charged Zn^{2+} and negatively charged Cl^- when electrolysed.

ion exchange
process whereby an ion in one compound is replaced by a different ion, of the same charge, from another compound. It is the basis of a type of chromatography in which the components of a mixture of ions in solution are separated according to the ease with which they will replace the ions on the polymer matrix through which they flow. The exchange of positively charged ions is called cation exchange; that of negatively charged ions is called anion exchange.

Ion exchange is used in commercial water softeners to exchange the dissolved ions responsible for the water's hardness with others that do not have this effect. For example, when hard water is passed over an ion-exchange resin, the dissolved calcium and magnesium ions are replaced by either sodium or hydrogen ions, so the hardness is removed.

The addition of washing soda crystals to hard water is also an example of ion exchange.

$$Na_2CO_{3\,(aq)} + CaSO_{4\,(aq)} \rightarrow CaCO_{3\,(s)} + Na_2SO_{4\,(aq)}$$

ionic bond or *electrovalent bond*
bond produced when atoms of one element donate electrons to atoms of another element, forming positively and negatively charged ions respectively. The attraction between the oppositely charged ions constitutes the bond. Sodium chloride (Na^+Cl^-) is a typical ionic compound.

Each ion has the electronic structure of a rare gas (see noble gas structure). The maximum number of electrons that can be gained is usually two.

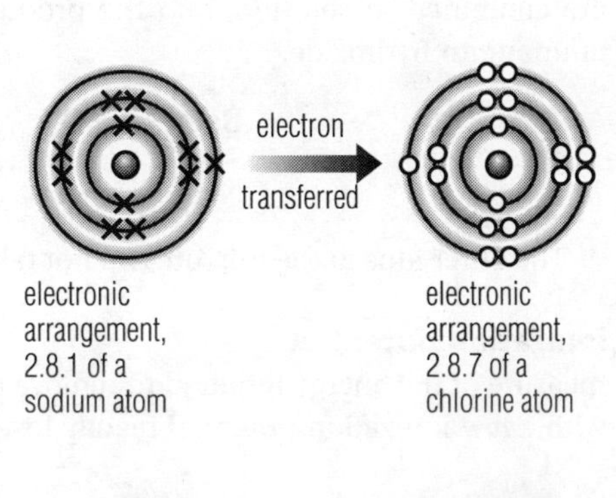

becomes a sodium ion, Na^+, with an electron arrangement 2.8

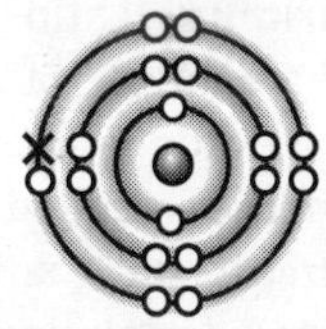

becomes a chloride ion, Cl^-, with an electron arrangement 2.8.8

The formation of an ionic bond between a sodium atom and a chlorine atom to form a molecule of sodium chloride. The sodium atom transfers an electron from its outer electron shell (becoming the positive ion Na^+) to the chlorine atom (which becomes the negative chloride ion Cl^-). The opposite charges mean that the ions are strongly attracted to each other. The formation of the bond means that each atom becomes more stable, having a full quota of electrons in its outer shell.

ionic compound

substance composed of oppositely charged ions. All salts, most bases, and some acids are examples of ionic compounds. They possess the following general properties: they are crystalline solids with a high melting point; are soluble in water and insoluble in organic solvents; and always conduct electricity when molten or in aqueous solution. A typical ionic compound is sodium chloride (Na^+Cl^-).

ionic equation

equation showing only those ions in a chemical reaction that actually undergo a change, either by combining together to form an insoluble salt or by combining together to form one or more molecular compounds. Examples are the precipitation of insoluble barium sulphate when barium and sulphate ions are combined in solution, and the production of ammonia and water from ammonium hydroxide.

$$Ba^{2+}_{(aq)} + SO_4{}^{2-}{}_{(aq)} \rightarrow BaSO_{4(s)}$$

$$NH_4{}^{+}{}_{(aq)} + OH^-{}_{(aq)} \rightarrow NH_{3(g)} + H_2O_{(l)}$$

The other ions in the mixtures do not take part and are called spectator ions.

ionization potential

measure of the energy required to remove an electron from an atom. Elements with a low ionization potential readily lose electrons to form cations.

iridium (Latin *iridis* 'rainbow')

hard, brittle, silver-white, metallic element, symbol Ir, atomic number 77, relative atomic mass 192.2. It is resistant to tarnish and corrosion. Iridium is one of the so-called platinum group of metals; it occurs in platinum ores and as a free metal (native metal) with osmium in osmiridium, a natural alloy that includes platinum, ruthenium, and rhodium.

It is alloyed with platinum for jewellery and used for watch bearings and in scientific instruments. It was named in 1804 by English chemist Smithson Tennant (1761–1815) for the iridescence of its salts in solution.

iron (Germanic *eis* 'strong')

hard, malleable and ductile, silver-grey, metallic element, symbol Fe (from Latin *ferrum*), atomic number 26, relative atomic mass 55.847. It is the fourth most abundant element (the second most abundant metal, after aluminium) in the Earth's crust. Iron occurs in concentrated deposits as the ores hematite (Fe_2O_3), spathic ore ($FeCO_3$), and magnetite (Fe_3O_4). It sometimes occurs as a free metal, occasionally as fragments of iron or iron–nickel meteorites.

Iron is the most common and most useful of all metals; it is strongly magnetic and is the basis for steel, an alloy with carbon and other elements. In electrical equipment it is used in permanent magnets and electromagnets, and forms the cores of transformers and magnetic amplifiers. It is noted for becoming oxidized (rusted) in moist air. In the human body, iron is an essential component of haemoglobin, the molecule in red blood cells that transports oxygen to all parts of the body. A deficiency in the diet causes a form of anaemia.

butane $CH_3(CH_2)_2CH_3$

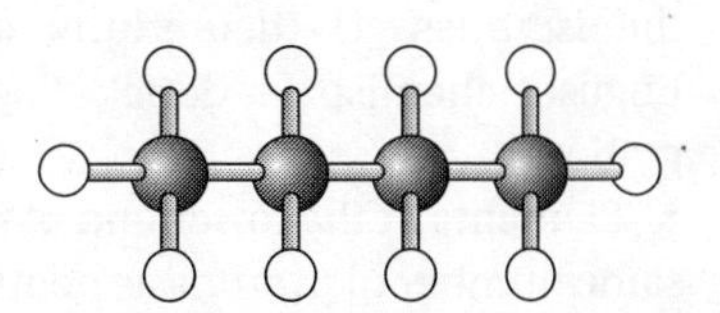

methyl propane $CH_3CH(CH_3)CH_3$

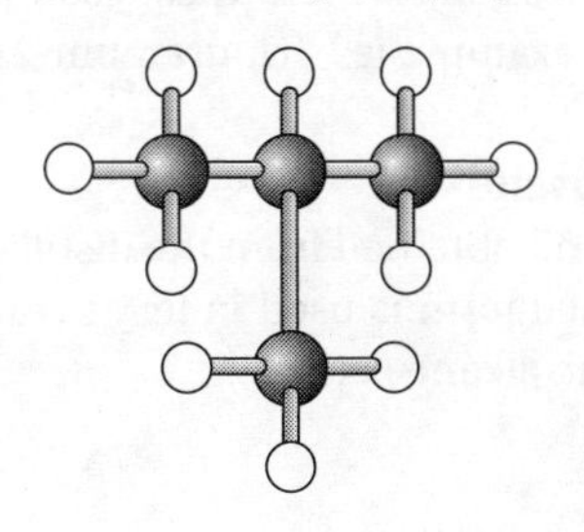

hydrogen atom
carbon atom
covalent bond

The chemicals butane and methyl propane are isomers. Each has the molecular formula C_4H_{10}, but with different spatial arrangements of atoms in their molecules.

isomer

chemical compound having the same molecular composition and mass as another, but with different physical or chemical properties owing to the different structural arrangement of its constituent atoms. For example, the organic compounds butane ($CH_3(CH_2)_2CH_3$) and methyl propane ($CH_3CH(CH_3)$ CH_3) are isomers, each possessing four carbon atoms and ten hydrogen atoms but differing in the way that these are arranged with respect to each other.

Structural isomers have obviously different constructions, but ***geometrical*** and ***optical isomers*** must be drawn or modelled in order to appreciate the difference in their three-dimensional arrangement.Geometrical isomers have a plane of symmetry and arise because of the restricted rotation of atoms around a bond; optical isomers are mirror images of each other. For instance, 1,1-dichloroethene (CH_2=CCl_2) and 1,2-dich-loroethene (CHCl=CHCl) are structural isomers, but there are two possible geometric isomers of the latter (depending on whether the chlorine atoms are on the same side or on opposite sides of the plane of the carbon–carbon double bond).

isomorphism

existence of substances of different chemical composition but with similar crystalline form.

isoprene

$CH_2CHC(CH_3)CH_2$ (technical name ***methylbutadiene***) colourless, volatile fluid obtained from petroleum and coal, used to make synthetic rubber.

isotope

one of two or more atoms that have the same atomic number (same number of protons), but which contain a different number of neutrons, thus differing in their atomic mass (see relative atomic mass). They may be stable or radioactive, naturally occurring, or synthesized. For example, hydrogen has the isotopes ^{2}H (deuterium) and ^{3}H (tritium). The term was coined by English chemist Frederick Soddy, pioneer researcher in atomic disintegration.

Elements at the lower end of the periodic table have atoms with roughly the same number of protons as neutrons. These elements are called ***stable isotopes.*** The stable isotopes of oxygen include ^{16}O, ^{17}O, and ^{18}O. Elements with high atomic mass numbers have many more neutrons than protons and are therefore less stable. It is these isotopes that are more prone to radioactive decay. One example is ^{238}U, uranium-238.

kerosene
thin oil obtained from the distillation of petroleum (formerly paraffin); a highly refined form is used in jet aircraft fuel. Kerosene is a mixture of hydrocarbons of the alkane series.

ketone
member of the group of organic compounds containing the carbonyl group (C=O) bonded to two atoms of carbon (instead of one carbon and one hydrogen as in aldehydes). Ketones are liquids or low-melting-point solids, slightly soluble in water.

An example is propanone (acetone, CH_3COCH_3), used as a solvent.

kinetics
branch of chemistry that investigates the rates of chemical reactions.

krypton (Greek *kryptos* 'hidden')
colourless, odourless, gaseous, nonmetallic element, symbol Kr, atomic number 36, relative atomic mass 83.80. It is grouped with the rare gases and was long believed not to enter into reactions, but it is now known to combine with fluorine under certain conditions; it remains inert to all other reagents. It is present in very small quantities in the air (about 114 parts per million). It is used chiefly in fluorescent lamps, lasers, and gas-filled electronic valves.

Krypton was discovered in 1898 in the residue from liquid air by British chemists William Ramsay and Morris Travers; the name refers to their difficulty in isolating it.

lactic acid or *2-hydroxypropanoic acid*
$CH_3CHOHCOOH$ organic acid, a colourless, almost odourless liquid, produced by certain bacteria during fermentation and by active muscle cells when they are exercised hard and are experiencing oxygen debt. An accumulation of lactic acid in the muscles may cause cramp. It occurs in yogurt, buttermilk, sour cream, poor wine, and certain plant extracts, and is used in food preservation and in the preparation of pharmaceuticals.

lanthanide

any of a series of 15 metallic elements (also known as rare earths) with atomic numbers 57 (lanthanum) to 71 (lutetium). One of its members, promethium, is radioactive. All occur in nature. Lanthanides are grouped because of their chemical similarities (most are trivalent, but some can be divalent or tetravalent), their properties differing only slightly with atomic number.

Lanthanides were called rare earths originally because they were not widespread and were difficult to identify and separate from their ores by their discoverers. The series is set out in a band in the periodic table of the elements, as are the actinides.

lanthanum (Greek *lanthanein* 'to be hidden')

soft, silvery, ductile and malleable, metallic element, symbol La, atomic number 57, relative atomic mass 138.91, the first of the lanthanide series. It is used in making alloys. It was named in 1839 by Swedish chemist Carl Mosander (1797–1858).

lawrencium

synthesized, radioactive, metallic element, the last of the actinide series, symbol Lr, atomic number 103, relative atomic mass 262. Its only known isotope, Lr-257, has a half-life of 4.3 seconds and was originally synthesized at the University of California at Berkeley in 1961 by bombarding californium with boron nuclei. The original symbol, Lw, was officially changed in 1963.

The element was named after Ernest Lawrence (1901–1958), the US inventor of the cyclotron.

lead

heavy, soft, malleable, grey, metallic element, symbol Pb (from Latin *plumbum*), atomic number 82, relative atomic mass 207.19. Usually found as an ore (most often as the sulphide galena), it occasionally occurs as a free metal (native metal), and is the final stable product of the decay of uranium. Lead is the softest and weakest of the commonly used metals, with a low melting point; it is a poor conductor of electricity and resists acid corrosion. As a cumulative poison, lead enters the body from lead water pipes, lead-based paints, and leaded petrol. (In humans, exposure to lead shortly after birth is associated with impaired mental health between the ages of two and four.) The metal is an effective shield against radiation and is used in batteries, glass, ceramics, and alloys such as pewter and solder.

Le Chatelier's principle or *Le Chatelier-Braun principle*

principle that if a change in conditions is imposed on a system in equilibrium, the system will react to counteract that change and restore the equilibrium. First stated in 1884 by French chemist Henri le Chatelier (1850–1936), it has been found to apply widely outside the field of chemistry.

lime or *quicklime*

CaO (technical name ***calcium oxide***) white powdery substance used in making mortar and cement. It is made commercially by heating calcium carbonate ($CaCO_3$), obtained from limestone or chalk, in a lime kiln. Quicklime readily absorbs water to become calcium hydroxide $Ca(OH)_2$, known as slaked lime, which is used to reduce soil acidity.

lipid

any of a large number of esters of fatty acids, commonly formed by the reaction of a fatty acid with glycerol (see glyceride). They are soluble in alcohol but not in water. Lipids are the chief constituents of plant and animal waxes, fats, and oils.

Phospholipids are lipids that also contain a phosphate group, usually linked to an organic base; they are major components of biological cell membranes.

lipophilic (Greek 'fat-loving')

term describing functional groups with an affinity for fats and oils.

liquefaction

process of converting a gas to a liquid, normally associated with low temperatures and high pressures (see condensation).

liquid

state of matter between a solid and a gas. A liquid forms a level surface and assumes the shape of its container. Its atoms do not occupy fixed positions as in a crystalline solid, nor do they have total freedom of movement as in a gas. Unlike a gas, a liquid is difficult to compress since pressure applied at one point is equally transmitted throughout (Pascal's principle). Hydraulics makes use of this property.

liquid air

air that has been cooled so much that it has liquefied. This happens at temperatures below about –196°C/–321°F. The various constituent gases, including nitrogen, oxygen, argon, and neon, can be separated from liquid air by the technique of fractionation.

Air is liquefied by the ***Linde process***, in which air is alternately compressed, cooled, and expanded, the expansion resulting each time in a considerable reduction in temperature. With the lower temperature the molecules move more slowly and occupy less space, so the air changes phase to become liquid.

lithium (Greek *lithos* 'stone')

soft, ductile, silver-white, metallic element, symbol Li, atomic number 3, relative atomic mass 6.941. It is one of the alkali metals, has a very low density (far less than most woods), and floats on water (specific gravity 0.57); it is the lightest of all metals. Lithium is used to harden alloys, and in batteries; its compounds are used in medicine to treat manic depression.

Lithium was named in 1818 by Swedish chemist Jöns Berzelius, having been discovered the previous year by his student Johan A Arfwedson (1792–1841). Berzelius named it after 'stone' because it is found in most igneous rocks and many mineral springs.

litmus
dye obtained from various lichens and used in chemistry as an indicator to test the acidic or alkaline nature of aqueous solutions; it turns red in the presence of acid, and blue in the presence of alkali.

lone pair
pair of electrons in the outermost shell of an atom that are not used in bonding. In certain circumstances, they will allow the atom to bond with atoms, ions, or molecules (such as boron trifluoride, BF_3) that are deficient in electrons, forming coordinate covalent (dative) bonds in which they provide both of the bonding electrons.

lutetium (Latin *Lutetia* 'Paris')
silver-white, metallic element, the last of the lanthanide series, symbol Lu, atomic number 71, relative atomic mass 174.97. It is used in the 'cracking', or breakdown, of petroleum and in other chemical processes. It was named by its discoverer, French chemist Georges Urbain, (1872–1938) after his native city.

magnesia
common name for magnesium oxide.

magnesium
lightweight, very ductile and malleable, silver-white, metallic element, symbol Mg, atomic number 12, relative atomic mass 24.305. It is one of the alkaline-earth metals, and the lightest of the commonly used metals. Magnesium silicate, carbonate, and chloride are widely distributed in nature. The metal is used in alloys, flares, and flash bulbs. It is a necessary trace element in the human diet, and green plants cannot grow without it since it is an essential constituent of the photosynthetic pigment chlorophyll ($C_{55}H_{72}MgN_4O_5$).

It was named after the ancient Greek city of Magnesia, near where it was first found. It was first recognized as an element by Scottish chemist Joseph Black 1755 and discovered in its oxide by English chemist Humphry Davy 1808. Pure magnesium was isolated in 1828 by French chemist Antoine-Alexandre-Brutus Bussy.

magnesium oxide or *magnesia*
MgO white powder or colourless crystals, formed when magnesium is burned in air or oxygen; a typical basic oxide. It is used to treat acidity of the stomach, and in some industrial processes; for example, as a lining brick in furnaces, because it is very stable when heated (refractory oxide).

malic acid or *hydroxysuccinic acid*
$HOOCCH_2CH(OH)COOH$ organic crystalline acid that can be extracted from apples, plums, cherries, grapes, and other fruits, but occurs in all living cells in smaller amounts, being one of the intermediates of the Krebs cycle.

maltase
enzyme found in plants and animals that breaks down the disaccharide maltose into glucose.

maltose
$C_{12}H_{22}O_{11}$ disaccharide sugar in which both monosaccharide units are glucose. It is produced by the enzymic hydrolysis of starch and is a major constituent of malt, produced in the early stages of beer and whisky manufacture.

manganese
hard, brittle, grey-white metallic element, symbol Mn, atomic number 25, relative atomic mass 54.9380. It resembles iron (and rusts), but it is not magnetic and is softer. It is used chiefly in making steel alloys, also alloys with aluminium and copper. It is used in fertilizers, paints, and industrial chemicals. It is a necessary trace element in human nutrition. The name is old, deriving from the French and Italian forms of Latin for *magnesia* (MgO), the white tasteless powder used as an antacid from ancient times.

mass action, law of
law stating that at a given temperature the rate at which a chemical reaction takes place is proportional to the product of the active masses of the reactants. The active mass is taken to be the molar concentration of each reactant.

mass number or *nucleon number*
sum (symbol *A*) of the numbers of protons and neutrons in the nucleus of an atom. It is used along with the atomic number (the number of protons) in nuclear notation: in symbols that represent nuclear isotopes, such as ${}^{14}_{6}C$, the lower number is the atomic number, and the upper number is the mass number.

meitnerium
synthesized radioactive element of the transactinide series, symbol Mt, atomic number 109, relative atomic mass 266. It was first produced in 1982 at the Laboratory for Heavy Ion Research in Darmstadt, Germany, by fusing bismuth and iron nuclei; it took a week to obtain a single new, fused nucleus. It was named in 1997 after the Austrian-born Swedish physicist Lise Meitner.

melting point
temperature at which a substance melts, or changes from solid to liquid form. A pure substance under standard conditions of pressure (usually one atmosphere) has a definite melting point. If heat is supplied to a solid at its melting point, the temperature does not change until the melting process is complete. The melting point of ice is 0°C or 32°F.

mendelevium

synthesized, radioactive metallic element of the actinide series, symbol Md, atomic number 101, relative atomic mass 258. It was first produced by bombardment of Es-253 with helium nuclei. Its longest-lived isotope, Md-258, has a half-life of about two months. The element is chemically similar to thulium. It was named by the US physicists at the University of California at Berkeley who first synthesized it in 1955 after the Russian chemist Dmitri Mendeleyev, who in 1869 devised the basis for the periodic table of the elements.

mercury or *quicksilver* (Latin *mercurius*)

heavy, silver-grey, metallic element, symbol Hg (from Latin *hydrargyrum*), atomic number 80, relative atomic mass 200.59. It is a dense, mobile liquid with a low melting point (–38.87°C/–37.96°F). Its chief source is the mineral cinnabar, HgS, but it sometimes occurs in nature as a free metal.

Its alloys with other metals are called amalgams (a silver–mercury amalgam is used in dentistry for filling cavities in teeth). Industrial uses include drugs and chemicals, mercury-vapour lamps, arc rectifiers, power-control switches, barometers, and thermometers. Mercury is a cumulative poison that can contaminate the food chain, and cause intestinal disturbance, kidney and brain damage, and birth defects in humans.

metal

any of a class of chemical elements with specific physical and chemical characteristics. Metallic elements compose about 75% of the 112 elements in the periodic table of the elements.

Physical properties include a sonorous tone when struck; good conduction of heat and electricity; opacity but good reflection of light; malleability, which enables them to be cold-worked and rolled into sheets; ductility, which permits them to be drawn into thin wires; and the possible emission of electrons when heated (thermionic effect) or when the surface is struck by light (photoelectric effect).

The majority of metals are found in nature in a combined form only, as compounds or mineral ores; about 16 of them also occur in the elemental form, as native metals. Their chemical properties are largely determined by the extent to which their atoms can lose one or more electrons and form positive ions (cations).

metallic bond

force of attraction operating in a metal that holds the atoms together. In the metal the valency electrons are able to move within the crystal and these electrons are said to be delocalized. Their movement creates short-lived, positively charged ions. The electrostatic attraction between the delocalized electrons and the ceaselessly forming ions constitutes the metallic bond.

metallic character
chemical properties associated with those elements classed as metals. These properties, which arise from the element's ability to lose electrons, are: the displacement of hydrogen from dilute acids; the formation of basic oxides; the formation of ionic chlorides; and their reducing reaction, as in the thermite process (see reduction).

In the periodic table of the elements, metallic character increases down any group and across a period from right to left.

metalloid or *semimetal*
any chemical element having some of but not all the properties of metals; metalloids are thus usually electrically semiconducting. They comprise the elements germanium, arsenic, antimony, and tellurium.

methanal or *formaldehyde*
HCHO gas at ordinary temperatures, condensing to a liquid at –21°C/–5.8°F. It has a powerful, penetrating smell. Dissolved in water, it is used as a biological preservative. It is used in the manufacture of plastics, dyes, foam (for example urea-formaldehyde foam, used in insulation), and in medicine.

methane
CH_4 the simplest hydrocarbon of the alkane series. Colourless, odourless, and lighter than air, it burns with a bluish flame and explodes when mixed with air or oxygen. It is the chief constituent of natural gas and also occurs in the explosive firedamp of coal mines. Methane emitted by rotting vegetation forms marsh gas, which may ignite by spontaneous combustion to produce the pale flame seen over marshland and known as will-o'-the-wisp.

methanoic acid or *formic acid*
HCOOH, colourless, slightly fuming liquid that freezes at 8°C/46.4°F and boils at 101°C/213.8°F. It occurs in stinging ants, nettles, sweat, and pine needles, and is used in dyeing, tanning, and electroplating.

methanol common name *methyl alcohol*
CH_3OH the simplest of the alcohols. It can be made by the dry distillation of wood (hence it is also known as wood alcohol), but is usually made from coal or natural gas. When pure, it is a colourless, flammable liquid with a pleasant odour, and is highly poisonous.

Methanol is used to produce methanal (formaldehyde, from which resins and plastics can be made), methyl-ter-butyl ether (MTB, a replacement for lead as an octane-booster in petrol), vinyl acetate (largely used in paint manufacture), and petrol.

methylated spirit
alcohol that has been rendered undrinkable, and is used for industrial purposes, as a fuel for spirit burners or a solvent. It is nevertheless drunk by some

individuals, resulting eventually in death. One of the poisonous substances in it is methanol, or methyl alcohol, and this gives it its name. (The 'alcohol' of alcoholic drinks is ethanol.)

mineral
naturally formed inorganic substance with a particular chemical composition and a regularly repeating internal structure. Either in their perfect crystalline form or otherwise, minerals are the constituents of rocks. In more general usage, a mineral is any substance economically valuable for mining (including coal and oil, despite their organic origins).

molarity
concentration of a solution expressed as the number of moles in grams of solute per cubic decimetre of solution.

molar volume
volume occupied by one mole (the molecular mass in grams) of any gas at standard temperature and pressure, equal to 2.24136×10^{-2} m^3.

mole
SI unit (symbol mol) of the amount of a substance. It is defined as the amount of a substance that contains as many elementary entities (atoms, molecules, and so on) as there are atoms in 12 g of the isotope carbon-12.

One mole of an element that exists as single atoms weighs as many grams as its atomic number (so one mole of carbon weighs 12 g), and it contains 6.022045×10^{23} atoms, which is Avogadro's number.

molecular formula
formula indicating the actual number of atoms of each element present in a single molecule of a chemical compound. This is determined by two pieces of information: the empirical formula and the relative molecular mass, which is determined experimentally.

molecular mass or *relative molecular mass*
mass of a molecule, calculated relative to one-twelfth the mass of an atom of carbon-12. It is found by adding the relative atomic masses of the atoms that make up the molecule.

molecule
smallest configuration of an element or compound that can exist independently. Hydrogen atoms, at room temperature, do not exist independently. They are bonded in pairs to form hydrogen molecules. A molecule of a compound consists of two or more different atoms bonded together. Molecules vary in size and complexity from the hydrogen molecule (H_2) to the large macromolecules of proteins. They may be held together by ionic bonds, in which the atoms gain

or lose electrons to form ions, or by covalent bonds, where electrons from each atom are shared in a new molecular orbital. Each compound is represented by a chemical symbol, indicating the elements into which it can be broken down and the number of each type of atom present. The symbolic representation of a molecule is known as its formula. For example, one molecule of the compound water, having two atoms of hydrogen and one atom of oxygen, is shown as H_2O.

molybdenum (Greek *molybdos* 'lead')
heavy, hard, lustrous, silver-white, metallic element, symbol Mo, atomic number 42, relative atomic mass 95.94. The chief ore is the mineral sulphide molybdenite. The element is highly resistant to heat and conducts electricity easily. It is used in alloys, often to harden steels. It is a necessary trace element in human nutrition. It was named in 1781 by Swedish chemist Karl Scheele, after its isolation by another Swedish chemist Peter Jacob Hjelm (1746–1813), for its resemblance to lead ore.

monomer
chemical compound composed of simple molecules from which polymers can be made. Under certain conditions the simple molecules (of the monomer) join together (polymerize) to form a very long chain molecule (macromolecule) called a polymer. For example, the polymerization of ethene (ethylene) monomers produces the polymer polyethene (polyethylene).

$$2nCH_2 = CH_2 \rightarrow (CH_2\text{–}CH_2\text{–}CH_2\text{–}\ CH_2)_n$$

naphtha
mixtures of hydrocarbons obtained by destructive distillation of petroleum, coal tar, and shale oil. It is a raw material for the petrochemical and plastics industries. The term was originally applied to naturally occurring liquid hydrocarbons.

naphthalene
$C_{10}H_8$ solid, white, shiny, aromatic hydrocarbon obtained from coal tar. The smell of moth balls is due to their naphthalene content. It is used in making indigo and certain azo dyes, as a mild disinfectant, and as a pesticide.

native metal or *free metal*
any of the metallic elements that occur in nature in the chemically uncombined or elemental form (in addition to any combined form). They include bismuth, cobalt, copper, gold, iridium, iron, lead, mercury, nickel, osmium, palladium, platinum, ruthenium, rhodium, tin, and silver. Some are commonly found in the free state, such as gold; others occur almost exclusively in the combined state, but under unusual conditions do occur as native metals, such as mercury. Examples of native nonmetals are carbon and sulphur.

neodymium
yellowish metallic element of the lanthanide series, symbol Nd, atomic number 60, relative atomic mass 144.24. Its rose-coloured salts are used in colouring glass, and neodymium is used in lasers. It was named in 1885 by Austrian chemist Carl von Welsbach (1858–1929), who fractionated it away from didymium (originally thought to be an element but actually a mixture of rare-earth metals consisting largely of neodymium, praesodymium, and cerium).

neon (Greek *neos* 'new')
colourless, odourless, nonmetallic, gaseous element, symbol Ne, atomic number 10, relative atomic mass 20.183. It is grouped with the rare gases, is nonreactive, and forms no compounds. It occurs in small quantities in the Earth's atmosphere.

Tubes containing neon are used in electric advertising signs, giving off a fiery red glow; it is also used in lasers. Neon was discovered by Scottish chemist William Ramsay and English chemist Morris Travers.

neptunium
silvery, radioactive metallic element of the actinide series, symbol Np, atomic number 93, relative atomic mass 237.048. It occurs in nature in minute amounts in pitchblende and other uranium ores, where it is produced from the decay of neutron-bombarded uranium in these ores. The longest-lived isotope, Np-237, has a half-life of 2.2 million years. The element can be produced by bombardment of U-238 with neutrons and is chemically highly reactive.

It was first synthesized in 1940 by US physicists E McMillan (1907–1991) and Edwin Philip Abelson (1913–), who named it after the planet Neptune (since it comes after uranium as the planet Neptune comes after Uranus). Neptunium was the first transuranic element to be synthesized.

neutralization
process occurring when the excess acid (or excess base) in a substance is reacted with added base (or added acid) so that the resulting substance is neither acidic nor basic.

In theory neutralization involves adding acid or base as required to achieve pH 7. When the colour of an indicator is used to test for neutralization, the final pH may differ from pH 7 depending upon the indicator used. It will also differ from 7 in reactions between strong acids and weak bases and weak acids and strong bases, because the salt formed will have acid or basic properties respectively.

neutron number
symbol N, the number of neutrons possessed by an atomic nucleus. Isotopes are atoms of the same element possessing different neutron numbers.

nickel
hard, malleable and ductile, silver-white metallic element, symbol Ni, atomic number 28, relative atomic mass 58.71. It occurs in igneous rocks and as a free

metal (native metal), occasionally occurring in fragments of iron-nickel meteorites. It is a component of the Earth's core, which is held to consist principally of iron with some nickel. It has a high melting point, low electrical and thermal conductivity, and can be magnetized. It does not tarnish and therefore is much used for alloys, electroplating, and for coinage.

It was discovered in 1751 by Swedish mineralogist Axel Cronstedt (1722–1765) and the name given as an abbreviated form of *kopparnickel*, Swedish 'false copper', since the ore in which it is found resembles copper but yields none.

niobium

soft, grey-white, somewhat ductile and malleable, metallic element, symbol Nb, atomic number 41, relative atomic mass 92.906. It occurs in nature with tantalum, which it resembles in chemical properties. It is used in making stainless steel and other alloys for jet engines and rockets and for making superconductor magnets.

Niobium was discovered in 1801 by the English chemist Charles Hatchett (1765–1847), who named it columbium (symbol Cb), a name that is still used in metallurgy. In 1844 it was renamed after Niobe by the German chemist Heinrich Rose (1795–1864) because of its similarity to tantalum (Niobe is the daughter of Tantalus in Greek mythology).

nitrate

salt or ester of nitric acid, containing the NO_3^- ion. Nitrates are used in explosives, in the chemical and pharmaceutical industries, in curing meat, and as fertilizers. They are the most water-soluble salts known and play a major part in the nitrogen cycle. Nitrates in the soil, whether naturally occurring or from inorganic or organic fertilizers, can be used by plants to make proteins and nucleic acids. However, runoff from fields can result in nitrate pollution.

nitric acid or *aqua fortis*

HNO_3 fuming acid obtained by the oxidation of ammonia or the action of sulphuric acid on potassium nitrate. It is a highly corrosive acid, dissolving most metals, and a strong oxidizing agent. It is used in the nitration and esterification of organic substances, and in the making of sulphuric acid, nitrates, explosives, plastics, and dyes.

nitrification

process that takes place in soil when bacteria oxidize ammonia, turning it into nitrates. Nitrates can be absorbed by the roots of plants, so this is a vital stage in the nitrogen cycle.

nitrite

salt or ester of nitrous acid, containing the nitrite ion (NO_2^-). Nitrites are used as preservatives (for example, to prevent the growth of botulism spores) and as colouring agents in cured meats such as bacon and sausages.

nitrogen (Greek *nitron* 'native soda', sodium or potassium nitrate)
colourless, odourless, tasteless, gaseous, nonmetallic element, symbol N, atomic number 7, relative atomic mass 14.0067. It forms almost 80% of the Earth's atmosphere by volume and is a constituent of all plant and animal tissues (in proteins and nucleic acids). Nitrogen is obtained for industrial use by the liquefaction and fractional distillation of air. Its compounds are used in the manufacture of foods, drugs, fertilizers, dyes, and explosives.

Nitrogen has been recognized as a plant nutrient, found in manures and other organic matter, from early times, long before the complex cycle of nitrogen fixation was understood. It was isolated in 1772 by the English chemist Daniel Rutherford (1749–1819) and named in 1790 by the French chemist Jean Chaptal (1756–1832).

nitrogen fixation
process by which nitrogen in the atmosphere is converted into nitrogenous compounds by the action of micro-organisms, such as cyanobacteria (blue-green algae) and bacteria, in conjunction with certain legumes. Several chemical processes duplicate nitrogen fixation to produce fertilizers.

nitrogen oxide
any chemical compound that contains only nitrogen and oxygen. All nitrogen oxides are gases. Nitrogen monoxide and nitrogen dioxide contribute to air pollution.

nitrogen monoxide
NO, or nitric oxide, is a colourless gas released when metallic copper reacts with concentrated nitric acid. It is also produced when nitrogen and oxygen combine at high temperature. On contact with air it is oxidized to nitrogen dioxide.

nitrogen dioxide
nitrogen(IV) oxide, NO_2, is a brown, acidic, pungent gas that is harmful if inhaled and contributes to the formation of acid rain, as it dissolves in water to form nitric acid. It is the most common of the nitrogen oxides and is obtained by heating most nitrate salts (for example lead(II) nitrate, $Pb(NO_3)_2$). If liquefied, it gives a colourless solution (N_2O_4). It has been used in rocket fuels.

nitroglycerine or *glyceryl trinitrate*
$C_3H_5(ONO_2)_3$ flammable, explosive oil produced by the action of nitric and sulphuric acids on glycerol. Although poisonous, it is used in cardiac medicine. It explodes with great violence if heated in a confined space and is used in the preparation of dynamite, cordite, and other high explosives.

nobelium
synthesized, radioactive, metallic element of the actinide series, symbol No, atomic number 102, relative atomic mass 259. It is synthesized by bombarding curium with carbon nuclei.

It was named in 1957 after the Nobel Institute in Stockholm, Sweden, where it was claimed to have been first synthesized. Later evaluations determined that this was in fact not so, as the successful 1958 synthesis at the University of California at Berkeley produced a different set of data. The name was not, however, challenged. In 1992 the International Unions for Pure and Applied Chemistry and Physics (IUPAC and IUPAP) gave credit to Russian scientists in Dubna for the discovery of nobelium.

noble gas
alternative name for rare gas.

noble gas structure
configuration of electrons in rare or noble gases (helium, neon, argon, krypton, xenon, and radon). This is characterized by full electron shells around the nucleus of an atom, which render the element stable. Any ion, produced by the gain or loss of electrons, that achieves an electronic configuration similar to one of the rare gases is said to have a noble gas structure.

nonmetal
one of a set of elements (around 20 in total) with certain physical and chemical properties opposite to those of metals. Nonmetals accept electrons (see electronegativity) and are sometimes called electronegative elements.

Their typical reactions are as follows.

with acids and alkalis
Nonmetals do not react with dilute acids but may react with alkalis.

$$2NaOH + Cl_2 \rightarrow NaCl + NaOCl + H_2O$$

with air or oxygen
They form acidic or neutral oxides.

$$S + O_2 \rightarrow SO_2$$

with chlorine
They react with chlorine gas to form covalent chlorides.

$$2P_{(s)} + 3Cl_2 \rightarrow 2PCl_3$$

with reducing agents
Nonmetals act as oxidizing agents.

$$2FeCl_2 + Cl_2 \rightarrow 2FeCl_3$$

nuclear notation
method used for labelling an atom according to the composition of its nucleus. The atoms or isotopes of a particular element are represented by the symbol ${}^{A}_{Z}X$ where A is the mass number of their nuclei, Z is their atomic number, and X is the chemical symbol for that element.

nucleic acid

complex organic acid made up of a long chain of nucleotides, present in the nucleus and sometimes the cytoplasm of the living cell. The two types, known as DNA (deoxyribonucleic acid) and RNA (ribonucleic acid), form the basis of heredity. The nucleotides are made up of a sugar (deoxyribose or ribose), a phosphate group, and one of four purine or pyrimidine bases. The order of the bases along the nucleic acid strand contains the genetic code.

nucleon

either a proton or a neutron, when present in the atomic nucleus. ***Nucleon number*** is an alternative name for the mass number of an atom.

nucleus

positively charged central part of an atom, which constitutes almost all its mass. Except for hydrogen nuclei, which have only one proton, nuclei are composed of both protons and neutrons. Surrounding the nuclei are electrons, of equal and opposite charge to that of the protons, thus giving the atom a neutral charge. The nucleus was discovered by the New Zealand-born British physicist Ernest Rutherford in 1911 as a result of experiments in firing alpha particles through very thin gold foil.

nylon

synthetic long-chain polymer similar in chemical structure to protein. Nylon was the first all-synthesized fibre, made from petroleum, natural gas, air, and water by the DuPont firm in 1938. It is used in the manufacture of moulded articles, textiles, and medical sutures. Nylon fibres are stronger and more elastic than silk and are relatively insensitive to moisture and mildew. Nylon is used for hosiery and woven goods, simulating other materials such as silks and furs; it is also used for carpets.

octane rating

numerical classification of petroleum fuels indicating their combustion characteristics. The efficient running of an internal combustion engine depends on the ignition of a petrol–air mixture at the correct time during the cycle of the engine. Higher-rated petrol burns faster than lower-rated fuels. The use of the correct grade must be matched to the engine.

orbital, atomic

region around the nucleus of an atom (or, in a molecule, around several nuclei) in which an electron is likely to be found. According to quantum theory, the position of an electron is uncertain; it may be found at any point. However, it is more likely to be found in some places than in others, and this pattern of probabilities makes up the orbital.

An atom or molecule has numerous orbitals, each of which has a fixed size and shape. An orbital is characterized by three numbers, called quantum numbers, representing its energy (and hence size), its angular momentum (and hence shape), and its orientation. Each orbital can be occupied by one or (if their spins are aligned in opposite directions) two electrons.

organic chemistry
branch of chemistry that deals with carbon compounds. Organic compounds form the chemical basis of life and are more abundant than inorganic compounds. In a typical organic compound, each carbon atom forms bonds covalently with each of its neighbouring carbon atoms in a chain or ring, and additionally with other atoms, commonly hydrogen, oxygen, nitrogen, or sulphur.

The basis of organic chemistry is the ability of carbon to form long chains of atoms, branching chains, rings, and other complex structures. Compounds containing only carbon and hydrogen are known as ***hydrocarbons***.

organic compound
class of compounds that contain carbon (carbonates, carbon monoxide, and carbon dioxide are excluded). The original distinction between organic and inorganic compounds was based on the belief that the molecules of living systems were unique, and could not be synthesized in the laboratory. Today it is routine to manufacture thousands of organic chemicals both in research and in the drug industry. Certain simple compounds of carbon, such as carbonates, oxides of carbon, carbon disulphide, and carbides are usually treated in inorganic chemistry.

osmium (Greek *osme* 'odour') or *ethanedioic acid*
hard, heavy, bluish-white, metallic element, symbol Os, atomic number 76, relative atomic mass 190.2. It is the densest of the elements, and is resistant to tarnish and corrosion. It occurs in platinum ores and as a free metal (see native metal) with iridium in a natural alloy called osmiridium, containing traces of platinum, ruthenium, and rhodium. Its uses include pen points and light-bulb filaments; like platinum, it is a useful catalyst.

It was discovered in 1803 and named in 1804 by English chemist Smithson Tennant (1761–1815) after the irritating smell of one of its oxides.

oxalic acid
$(COOH)_2.2H_2O$ white, poisonous solid, soluble in water, alcohol, and ether. Oxalic acid is found in rhubarb leaves, and its salts (oxalates) occur in wood sorrel (genus *Oxalis*, family Oxalidaceae) and other plants. It also occurs naturally in human body cells. It is used in the leather and textile industries, in dyeing and bleaching, ink manufacture, metal polishes, and for removing rust and ink stains.

oxidation
loss of electrons, gain of oxygen, or loss of hydrogen by an atom, ion, or molecule during a chemical reaction. Oxidation may be brought about by reaction with another compound (oxidizing agent), which simultaneously undergoes reduction, or electrically at the anode (positive electrode) of an electrolytic cell.

oxidation number

Roman numeral often seen in a chemical name, indicating the valency of the element immediately before the number. Examples are lead(II) nitrate, manganese(IV) oxide (manganese dioxide), and potassium manganate(VII) (potassium permanganate).

oxide

compound of oxygen and another element, frequently produced by burning the element or a compound of it in air or oxygen.

Oxides of metals are normally bases and will react with an acid to produce a salt in which the metal forms the cation (positive ion). Some of them will also react with a strong alkali to produce a salt in which the metal is part of a complex anion (negative ion; see amphoteric). Most oxides of nonmetals are acidic (dissolve in water to form an acid). Some oxides display no pronounced acidic or basic properties.

oxygen (Greek *oxys* 'acid'; *genes* 'forming')

colourless, odourless, tasteless, nonmetallic, gaseous element, symbol O, atomic number 8, relative atomic mass 15.9994. It is the most abundant element in the Earth's crust (almost 50% by mass), forms about 21% by volume of the atmosphere, and is present in combined form in water and many other substances. Oxygen is a by-product of photosynthesis and the basis for respiration in plants and animals.

Oxygen is very reactive and combines with all other elements except the rare gases and fluorine. It is present in carbon dioxide, silicon dioxide (quartz), iron ore, calcium carbonate (limestone). In nature it exists as a molecule composed of two atoms (O_2); single atoms of oxygen are very short-lived owing to their reactivity. They can be produced in electric sparks and by the Sun's ultraviolet radiation in the upper atmosphere, where they rapidly combine with molecular oxygen to form ozone (an allotrope of oxygen).

Oxygen is obtained for industrial use by the fractional distillation of liquid air, by the electrolysis of water, or by heating manganese (IV) oxide with potassium chlorate. It is essential for combustion, and is used with ethyne (acetylene) in high-temperature oxyacetylene welding and cutting torches.

The element was first identified by English chemist Joseph Priestley in 1774 and independently in the same year by Swedish chemist Karl Scheele. It was named by French chemist Antoine Lavoisier in 1777.

ozone

O_3 highly reactive pale-blue gas with a penetrating odour. Ozone is an allotrope of oxygen (see allotropy), made up of three atoms of oxygen. It is formed when the molecule of the stable form of oxygen (O_2) is split by ultraviolet radiation or electrical discharge. It forms the ozone layer in the upper atmosphere, which protects life on Earth from ultraviolet rays, a cause of skin cancer.

palladium

lightweight, ductile and malleable, silver-white, metallic element, symbol Pd, atomic number 46, relative atomic mass 106.4. It is one of the so-called platinum group of metals, and is resistant to tarnish and corrosion. It often occurs in nature as a free metal (see native metal) in a natural alloy with platinum. Palladium is used as a catalyst, in alloys of gold (to make white gold) and silver, in electroplating, and in dentistry. It was discovered in 1803 by British physicist William Wollaston (1766–1828), and named after the asteroid Pallas (found 1802).

pentanol

$C_5H_{11}OH$ (common name ***amyl alcohol***) clear, colourless, oily liquid, usually having a characteristic choking odour. It is obtained by the fermentation of starches and from the distillation of petroleum. There are eight possible isomers.

peptide

molecule comprising two or more amino acid molecules (not necessarily different) joined by ***peptide bonds***, whereby the acid group of one acid is linked to the amino group of the other (–CO.NHN). The number of amino acid molecules in the peptide is indicated by referring to it as a di-, tri-, or polypeptide (two, three, or many amino acids).

Proteins are built up of interacting polypeptide chains with various types of bonds occurring between the chains. Incomplete hydrolysis (splitting up) of a protein yields a mixture of peptides, examination of which helps to determine the sequence in which the amino acids occur within the protein.

peptide bond

bond that joins two peptides together within a protein. The carboxyl (–COOH) group on one amino acid reacts with the amino ($-NH_2$) group on another amino acid to form a peptide bond (–CO–NH–) with the elimination of water. Peptide bonds are broken by hydrolytic enzymes called peptidases.

period

horizontal row of elements in the periodic table. There is a gradation of properties along each period, from metallic (group I, the alkali metals) to nonmetallic (group VII, the halogens).

periodic table of the elements

table in which the elements are arranged in order of their atomic number. The table summarizes the major properties of the elements and enables predictions to be made about their behaviour.

There are striking similarities in the chemical properties of the elements in each of the groups (vertical columns), which are numbered I–VII and 0 to reflect the number of electrons in the outermost unfilled shell and hence the maximum valency. Reactivity increases down the group. A gradation (trend) of properties may be traced along the horizontal rows (called ***periods***). Metallic character increases across a period from right to left, and down a group. A large block of elements, between groups II and III, contains the transition elements, characterized by displaying more than one valence state. *See table on pages 230–1.*

petrol

mixture of hydrocarbons derived from petroleum, mainly used as a fuel for internal-combustion engines. It is colourless and highly volatile. ***Leaded petrol*** contains antiknock (a mixture of tetraethyl lead and dibromoethane), which improves the combustion of petrol and the performance of a car engine. The lead from the exhaust fumes enters the atmosphere, mostly as simple lead compounds. There is strong evidence that it can act as a nerve poison on young children and cause mental impairment. This has prompted a gradual switch to the use of ***unleaded petrol*** in the UK.

petroleum or *crude oil*

natural mineral oil, a thick greenish-brown flammable liquid found underground in permeable rocks. Petroleum consists of hydrocarbons mixed with oxygen, sulphur, nitrogen, and other elements in varying proportions. It is thought to be derived from ancient organic material that has been converted by, first, bacterial action, then heat, and pressure (but its origin may be chemical also).

From crude petroleum, various products are made by distillation and other processes; for example, fuel oil, petrol, kerosene, diesel, and lubricating oil. Petroleum products and chemicals are used in large quantities in the manufacture of detergents, artificial fibres, plastics, insecticides, fertilizers, pharmaceuticals, toiletries, and synthetic rubber.

pH

scale from 0 to 14 for measuring acidity or alkalinity. A pH of 7 indicates neutrality, below 7 is acid, while above 7 is alkaline. Strong acids, such as those used in car batteries, have a pH of about 2; strong alkalis such as sodium hydroxide are pH 13. Acidic fruits such as citrus fruits are about pH 4. Fertile soils have a pH of about 6.5 to 7.0, while weak alkalis such as soap are 9 to 10. *See illustration on page 232.*

phase

physical state of matter: for example, ice and liquid water are different phases of water; a mixture of the two is termed a two-phase system.

phenol

member of a group of aromatic chemical compounds with weakly acidic properties, which are characterized by a hydroxyl (OH) group attached directly to an aromatic ring. The simplest of the phenols, derived from benzene, is also

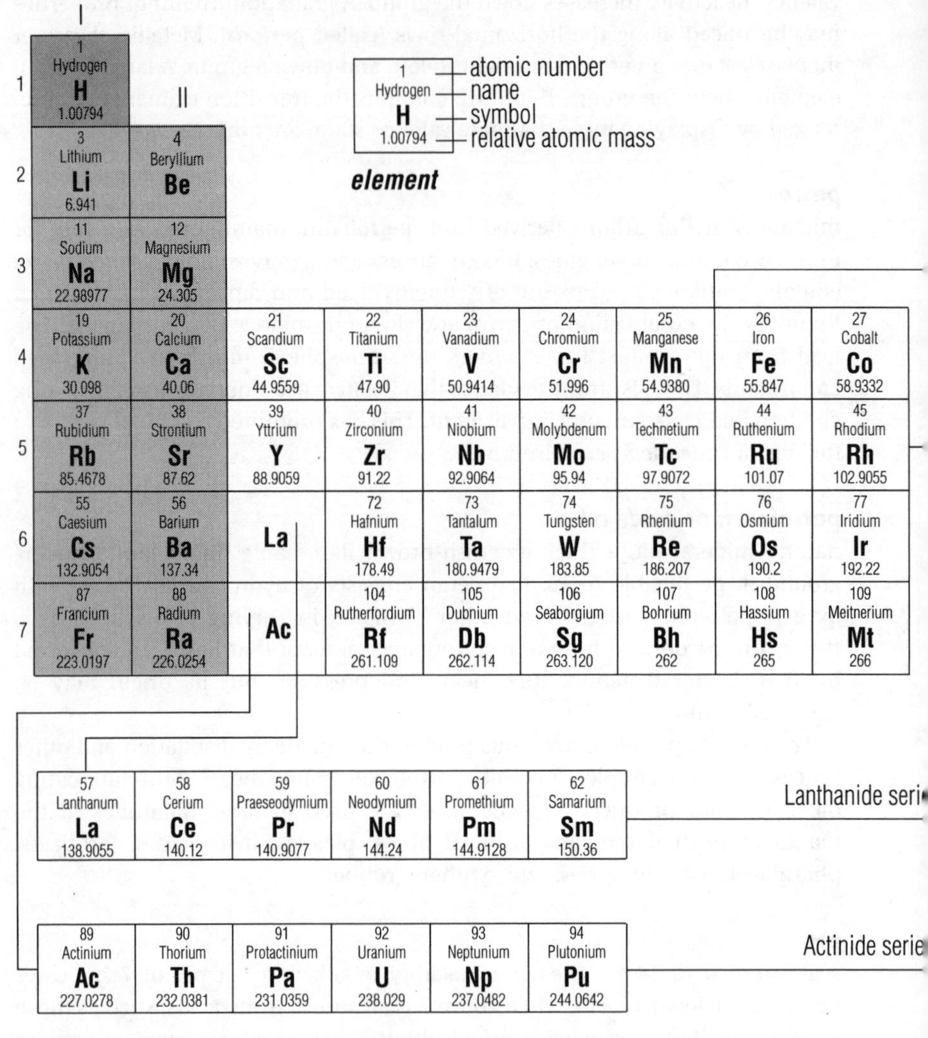

The periodic table of the elements arranges the elements into horizontal rows (called periods) and vertical columns (called groups) according to their atomic numbers.

known as phenol and has the formula C_6H_5OH. It is sometimes called ***carbolic acid*** and can be extracted from coal tar.

Pure phenol consists of colourless, needle-shaped crystals, which take up moisture from the atmosphere. It has a strong and characteristic smell and was once used as an antiseptic. It is, however, toxic by absorption through the skin.

phosphate
salt or ester of phosphoric acid. Incomplete neutralization of phosphoric acid gives rise to acid phosphates (see acid salt and buffer). Phosphates are used as

			III	IV	V	VI	VII	0
								2 Helium **He** 4002.60
			5 Boron **B** 10.81	6 Carbon **C** 12.011	7 Nitrogen **N** 14.0067	8 Oxygen **O** 15.9994	9 Fluorine **F** 18.99840	10 Neon **Ne** 20.179
			13 Aluminium **Al** 26.98154	14 Silicon **Si** 28.086	15 Phosphorus **P** 30.9738	16 Sulphur **S** 32.06	17 Chlorine **Cl** 35.453	18 Argon **Ar** 39.948
28 Nickel **Ni** 58.70	29 Copper **Cu** 63.546	30 Zinc **Zn** 65.38	31 Gallium **Ga** 69.72	32 Germanium **Ge** 72.59	33 Arsenic **As** 74.9216	34 Selenium **Se** 78.96	35 Bromine **Br** 79.904	36 Krypton **Kr** 83.80
46 Palladium **Pd** 106.4	47 Silver **Ag** 107.868	48 Cadmium **Cd** 112.40	49 Indium **In** 114.82	50 Tin **Sn** 118.69	51 Antimony **Sb** 121.75	52 Tellurium **Te** 127.75	53 Iodine **I** 126.9045	54 Xenon **Xe** 131.30
78 Platinum **Pt** 195.09	79 Gold **Au** 196.9665	80 Mercury **Hg** 200.59	81 Thallium **Tl** 204.37	82 Lead **Pb** 207.37	83 Bismuth **Bi** 207.2	84 Polonium **Po** 210	85 Astatine **At** 211	86 Radon **Rn** 222.0176
110 Ununnilium **Uun** 269	111 Unununium **Uuu** 272							

63 Europium **Eu** 151.96	64 Gadolinium **Gd** 157.25	65 Terbium **Tb** 158.9254	66 Dysprosium **Dy** 162.50	67 Holmium **Ho** 164.9304	68 Erbium **Er** 167.26	69 Thulium **Tm** 168.9342	70 Ytterbium **Yb** 173.04	71 Lutetium **Lu** 174.97
95 Americium **Am** 243.0614	96 Curium **Cm** 247.0703	97 Berkelium **Bk** 247	98 Californium **Cf** 251.0786	99 Einsteinium **Es** 252.0828	100 Fermium **Fm** 257.0951	101 Mendelevium **Md** 258.0986	102 Nobelium **No** 259.1009	103 Lawrencium **Lr** 260.1054

The elements in a group or column all have similar properties – for example, all the elements in the far right-hand column are inert gases.

fertilizers, and are required for the development of healthy root systems. They are involved in many biochemical processes, often as part of complex molecules, such as ATP.

phosphoric acid

acid derived from phosphorus and oxygen. Its commonest form (H_3PO_4) is also known as orthophosphoric acid, and is produced by the action of phosphorus pentoxide (P_2O_5) on water. It is used in rust removers and for rust-proofing iron and steel.

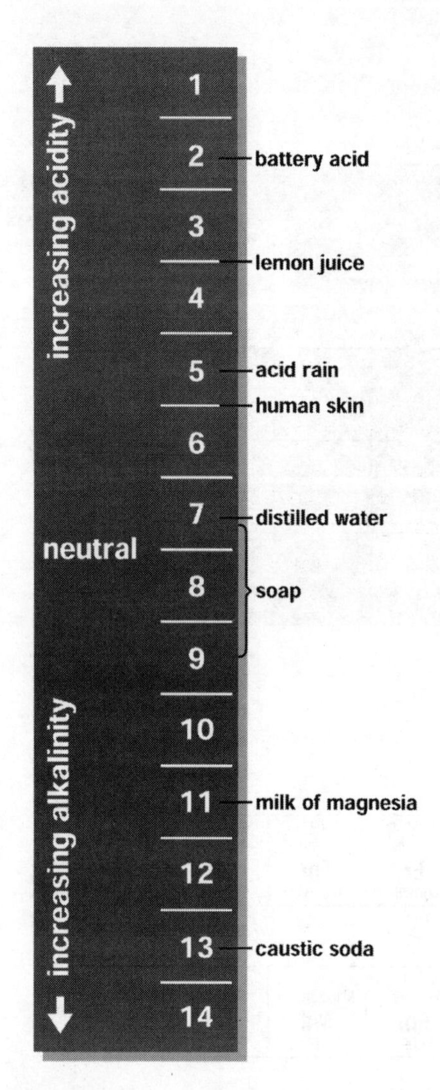

pH levels of some common substances. The lower the pH, the more acidic the substance; the higher the pH, the more alkaline the substance.

phosphorus (Greek *phosphoros* 'bearer of light')
highly reactive, nonmetallic element, symbol P, atomic number 15, relative atomic mass 30.9738. It occurs in nature as phosphates (commonly in the form of the mineral apatite), and is essential to plant and animal life. Compounds of phosphorus are used in fertilizers, various organic chemicals, for matches and fireworks, and in glass and steel.
Phosphorus was first identified in 1674 by German alchemist Hennig Brand (born *c.* 1630), who prepared it from urine. The element has three allotropic forms: a black powder; a white-yellow, waxy solid that ignites spontaneously in air to form the poisonous gas phosphorus pentoxide; and a red-brown powder that neither ignites spontaneously nor is poisonous.

photochemical reaction
any chemical reaction in which light is produced or light initiates the reaction. Light can initiate reactions by exciting atoms or molecules and making them more reactive: the light energy becomes converted to chemical energy. Many photochemical reactions set up a chain reaction and produce free radicals.

physical chemistry
branch of chemistry concerned with examining the relationships between the chemical compositions of substances and the physical properties that they display. Most chemical reactions exhibit some physical phenomenon (change of state, temperature, pressure, or volume, or the use or production of electricity), and the measurement and study of such phenomena has led to many chemical theories and laws.

pitchblende or *uraninite*
brownish-black mineral, the major uranium ore, consisting mainly of uranium oxide (UO_2). It also contains some lead (the final, stable product of uranium decay) and variable amounts of most of the naturally occurring radioactive elements, which are products of either the decay or the fissioning of uranium isotopes. The uranium yield is 50–80%; it is also a source of radium, polonium, and actinium. Pitchblende was first studied by Pierre and Marie Curie, who found radium and polonium in its residues in 1898.

plastic
any of the stable synthetic materials that are fluid at some stage in their manufacture, when they can be shaped, and that later set to rigid or semi-rigid solids. Plastics today are chiefly derived from petroleum. Most are polymers, made up of long chains of identical molecules.

Processed by extrusion, injection-moulding, vacuum-forming, and compression, plastics emerge in consistencies ranging from hard and inflexible to soft and rubbery. They replace an increasing number of natural substances, being lightweight, easy to clean, durable, and capable of being rendered very strong – for example, by the addition of carbon fibres – for building aircraft and other engineering projects.

platinum (Spanish *platina* 'little silver' (*plata* 'silver'))
heavy, soft, silver-white, malleable and ductile, metallic element, symbol Pt, atomic number 78, relative atomic mass 195.09. It is the first of a group of six metallic elements (platinum, osmium, iridium, rhodium, ruthenium, and palladium) that possess similar traits, such as resistance to tarnish, corrosion, and attack by acid, and that often occur as free metals (native metals). They often occur in natural alloys with each other, the commonest of which is osmiridium. Both pure and as an alloy, platinum is used in dentistry, jewellery, and as a catalyst.

plutonium
silvery-white, radioactive, metallic element of the actinide series, symbol Pu, atomic number 94, relative atomic mass 239.13. It occurs in nature in minute quantities in pitchblende and other ores, but is produced in quantity only synthetically. It has six allotropic forms (see allotropy) and is one of three fissile elements (elements capable of splitting into other elements – the others are thorium and uranium). Plutonium dioxide, PuO_2, a yellow crystalline solid, is the compound most widely used in the nuclear industry. It was believed to be inert until US researchers discovered in 1999 that it reacts very slowly with oxygen and water to form a previously unknown green crystalline compound that is soluble in water.

polyethylene or *polyethene*
polymer of the gas ethylene (technically called ethene, C_2H_4). It is a tough, white, translucent, waxy thermoplastic (which means it can be repeatedly soft-

ened by heating). It is used for packaging, bottles, toys, wood preservation, electric cable, pipes, and tubing.

Polyethylene is produced in two forms: low-density polyethylene, made by high-pressure polymerization of ethylene gas, and high-density polyethylene, which is made at lower pressure by using catalysts. This form, first made in 1953 by German chemist Karl Ziegler, is more rigid at low temperatures and softer at higher temperatures than the low-density type. Polyethylene was first made in the 1930s at very high temperatures by ICI.

polymer
compound made up of a large long-chain or branching matrix composed of many repeated simple units (***monomers***) linked together by polymerization. There are many polymers, both natural (cellulose, chitin, lignin, rubber) and synthetic (polyethylene and nylon, types of plastic). Synthetic polymers belong to two groups: thermosoftening and thermosetting. The size of the polymer matrix is determined by the amount of monomer used; it therefore does not form a molecule of constant molecular size or mass.

polymerization
chemical union of two or more (usually small) molecules of the same kind to form a new compound. ***Addition polymerization*** produces simple multiples of the same compound. ***Condensation polymerization*** joins molecules together with the elimination of water or another small molecule.

Addition polymerization uses only a single monomer (basic molecule); condensation polymerization may involve two or more different monomers (***co-polymerization***).

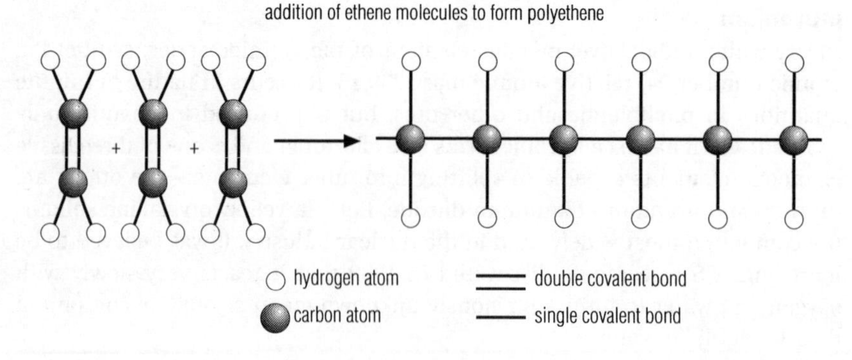

In polymerization, small molecules (monomers) join together to make large molecules (polymers). In the polymerization of ethene to polyethene, electrons are transferred from the carbon–carbon double bond of the ethene molecule, allowing the molecules to join together as a long chain of carbon–carbon single bonds.

polypropylene
plastic made by the polymerization, or linking together, of propene molecules (CH_2=CH-CH_3). It is used as a moulding material.

polysaccharide
long-chain carbohydrate made up of hundreds or thousands of linked simple sugars (monosaccharides) such as glucose and closely related molecules. The polysaccharides are natural polymers. They either act as energy-rich food stores in plants (starch) and animals (glycogen), or have structural roles in the plant cell wall (cellulose, pectin) or the tough outer skeleton of insects and similar creatures (chitin). See also carbohydrate.

Polythene
trade name for a variety of polyethene (polyethylene).

polyurethane
polymer made from the monomer urethane. It is a thermoset plastic, used in liquid form as a paint or varnish, and in foam form for upholstery and in lining materials (where it may be a fire hazard).

potash
general name for any potassium-containing mineral, most often applied to potassium carbonate (K_2CO_3) or potassium hydroxide (KOH). Potassium carbonate, originally made by roasting plants to ashes in earthenware pots, is commercially produced from the mineral sylvite (potassium chloride, KCl) and is used mainly in making artificial fertilizers, glass, and soap. The potassium content of soils and fertilizers is also commonly expressed as potash, although in this case it usually refers to potassium oxide (K_2O).

potassium (Dutch *potassa* 'potash')
soft, waxlike, silver-white, metallic element, symbol K (Latin *kalium*), atomic number 19, relative atomic mass 39.0983. It is one of the alkali metals and has a very low density – it floats on water, and is the second lightest metal (after lithium). It oxidizes rapidly when exposed to air and reacts violently with water. Of great abundance in the Earth's crust, it is widely distributed with other elements and found in salt and mineral deposits in the form of potassium aluminium silicates. The element was discovered and named in 1807 by English chemist Humphry Davy, who isolated it from potash in the first instance of a metal being isolated by electrolysis.

praseodymium (Greek *prasios* 'leek-green' + *didymos* 'twin')
silver-white, malleable, metallic element of the lanthanide series, symbol Pr, atomic number 59, relative atomic mass 140.907. It occurs in nature in the minerals monzanite and bastnaesite, and its green salts are used to colour glass and ceramics. It was named in 1885 by Austrian chemist Carl von Welsbach (1858–1929). He fractionated it from dydymium (originally thought to be an

element but actually a mixture of rare-earth metals consisting largely of neodymium, praseodymium, and cerium) and named it for its green salts and spectroscopic line.

precipitation
formation of an insoluble solid in a liquid as a result of a reaction within the liquid between two or more soluble substances. If the solid settles, it forms a ***precipitate***; if the particles of solid are very small, they will remain in suspension, forming a ***colloidal precipitate*** (see colloid).

promethium
radioactive, metallic element of the lanthanide series, symbol Pm, atomic number 61, relative atomic mass 145. It occurs in nature only in minute amounts, produced as a fission product/by-product of uranium in pitchblende and other uranium ores; for a long time it was considered not to occur in nature. The longest-lived isotope has a half-life of slightly more than 20 years.

Promethium is synthesized by neutron bombardment of neodymium and is a product of the fission of uranium, thorium, or plutonium; it can be isolated in large amounts from the fission-product debris of uranium fuel in nuclear reactors. It is used in phosphorescent paints and as an X-ray source.

propane
C_3H_8 gaseous hydrocarbon of the alkane series, found in petroleum and used as fuel and as a refrigerant.

propanol or *propyl alcohol*
third member of the homologous series of alcohols. Propanol is usually a mixture of two isomeric compounds (see isomer): propan-1-ol (CH_3CH_2 CH_2OH) and propan-2-ol ($CH_3CHOHCH_3$). Both are colourless liquids that can be mixed with water and are used in perfumery.

propanone
CH_3COCH_3 (common name ***acetone***) colourless flammable liquid used extensively as a solvent, as in nail-varnish remover, and for making acrylic plastics. It boils at 56.5°C/133.7°F, mixes with water in all proportions, and has a characteristic odour.

propene
$CH_3CH{=}CH_2$ (common name ***propylene***) second member of the alkene series of hydrocarbons. A colourless, flammable gas, it is widely used by industry to make organic chemicals, including polypropylene plastics.

protactinium (Latin *protos* 'before' + *aktis* 'first ray')
silver-grey, radioactive, metallic element of the actinide series, symbol Pa, atomic number 91, relative atomic mass 231.036. It occurs in nature in very

small quantities, in pitchblende and other uranium ores. It has 14 known isotopes; the longest-lived, Pa-231, has a half-life of 32,480 years.

protein

complex, biologically important substance composed of amino acids joined by peptide bonds. Proteins are essential to all living organisms. As enzymes they regulate all aspects of metabolism. Structural proteins such as ***keratin*** and ***collagen*** make up the skin, claws, bones, tendons, and ligaments; ***muscle*** proteins produce movement; ***haemoglobin*** transports oxygen; and ***membrane*** proteins regulate the movement of substances into and out of cells. For humans, protein is an essential part of the diet, and is found in greatest quantity in soy beans and other grain legumes, meat, eggs, and cheese.

Other types of bond, such as sulphur–sulphur bonds, hydrogen bonds, and cation bridges between acid sites, are responsible for creating the protein's characteristic three-dimensional structure, which may be fibrous, globular, or pleated. Protein provides 4 kcal of energy per gram (60 g per day is required).

proton number

alternative name for atomic number.

qualitative analysis

procedure for determining the identity of the component(s) of a single substance or mixture. A series of simple reactions and tests can be carried out on a compound to determine the elements present.

quantitative analysis

procedure for determining the precise amount of a known component present in a single substance or mixture. A known amount of the substance is subjected to particular procedures.

Gravimetric analysis determines the mass of each constituent present; ***volumetric analysis*** determines the concentration of a solution by titration against a solution of known concentration.

quicksilver

another name for the element mercury.

radical

group of atoms forming part of a molecule, which acts as a unit and takes part in chemical reactions without disintegration, yet often cannot exist alone for any length of time; for example, the methyl radical $-CH_3$, or the carboxyl radical $-COOH$.

radioactive decay

process of disintegration undergone by the nuclei of radioactive elements, such as radium and various isotopes of uranium and the transuranic elements. This changes the element's atomic number, thus transmuting one element into another, and is accompanied by the emission of radiation. Alpha and beta decay are the most common forms.

In ***alpha decay*** (the loss of a helium nucleus – two protons and two neutrons) the atomic number decreases by two and a new nucleus is formed, for example, an atom of uranium isotope of mass 238, on emitting an alpha particle, becomes an atom of thorium, mass 234. In ***beta decay*** the loss of an electron from an atom is accomplished by the transformation of a neutron into a proton, thus resulting in an increase in the atomic number of one. For example, the decay of the carbon-14 isotope results in the formation of an atom of nitrogen (mass 14, atomic number 7) and the emission of an electron. Gamma emission usually occurs as part of alpha or beta emission. In gamma emission high-speed electromagnetic radiation is emitted from the nucleus, making it more stable during the loss of an alpha or beta particle. Certain lighter artificially created isotopes also undergo radioactive decay. The associated radiation consists of alpha rays, beta rays, or gamma rays (or a combination of these), and it takes place at a constant rate expressed as a specific half-life, which is the time taken for half of any mass of that particular isotope to decay completely. Less commonly occurring decay forms include heavy-ion emission, electron capture, and spontaneous fission (in each of these the atomic number decreases). The original nuclide is known as the parent substance, and the product is a daughter nuclide (which may or may not be radioactive). The final product in all modes of decay is a stable element.

radioactivity

spontaneous alteration of the nuclei of radioactive atoms, accompanied by the emission of radiation. It is the property exhibited by the radioactive isotopes of stable elements and all isotopes of radioactive elements, and can be either natural or induced. See radioactive decay.

radiochemistry

chemical study of radioactive isotopes and their compounds (whether produced from naturally radioactive or irradiated materials) and their use in the study of other chemical processes.

When such isotopes are used in labelled compounds, they enable the biochemical and physiological functioning of parts of the living body to be observed. They can help in the testing of new drugs, showing where the drug goes in the body and how long it stays there. They are also useful in diagnosis – for example cancer, fetal abnormalities, and heart disease.

radium (Latin *radius* 'ray')

white, radioactive, metallic element, symbol Ra, atomic number 88, relative atomic mass 226.02. It is one of the alkaline-earth metals, found in nature in pitchblende and other uranium ores. Of the 16 isotopes, the commonest, Ra-226, has a half-life of 1,620 years. The element was discovered and named in

1898 by Pierre and Marie Curie, who were investigating the residues of pitchblende.

Radium decays in successive steps to produce radon (a gas), polonium, and finally a stable isotope of lead. The isotope Ra-223 decays through the uncommon mode of heavy-ion emission, giving off carbon-14 and transmuting directly to lead. Because radium luminesces, it was formerly used in paints that glowed in the dark; when the hazards of radioactivity became known its use was abandoned, but factory and dump sites remain contaminated and many former workers and neighbours contracted fatal cancers.

radon

colourless, odourless, gaseous, radioactive, nonmetallic element, symbol Rn, atomic number 86, relative atomic mass 222. It is grouped with the rare gases and was formerly considered nonreactive, but is now known to form some compounds with fluorine. Of the 20 known isotopes, only three occur in nature; the longest half-life is 3.82 days (Rn-222).

Radon is the densest gas known and occurs in small amounts in spring water, streams, and the air, being formed from the natural radioactive decay of radium. Ernest Rutherford discovered the isotope Rn-220 in 1899, and Friedrich Dorn in 1900; after several other chemists discovered additional isotopes, William Ramsay and R W Whytlaw-Gray isolated the element, which they named niton in 1908. The name radon was adopted in the 1920s.

rare-earth element

alternative name for lanthanide.

rare gas or *noble gas* or *inert gas*

any of a group of six elements (helium, neon, argon, krypton, xenon, and radon), so named because they were originally thought not to enter into any chemical reactions. This is now known to be incorrect: in 1962, xenon was made to combine with fluorine, and since then, compounds of argon, krypton, and radon with fluorine and/or oxygen have been described.

The extreme unreactivity of the rare gases is due to the stability of their electronic structure. All the electron shells (energy levels) of inert gas atoms are full and, except for helium, they all have eight electrons in their outermost (valency) shell. The apparent stability of this electronic arrangement led to the formulation of the octet rule to explain the different types of chemical bond found in simple compounds.

reaction

coming together of two or more atoms, ions, or molecules with the result that a chemical change takes place; that is, a change that occurs when two or more substances interact with each other, resulting in the production of different substances with different chemical compositions. The nature of the reaction is portrayed by a chemical equation.

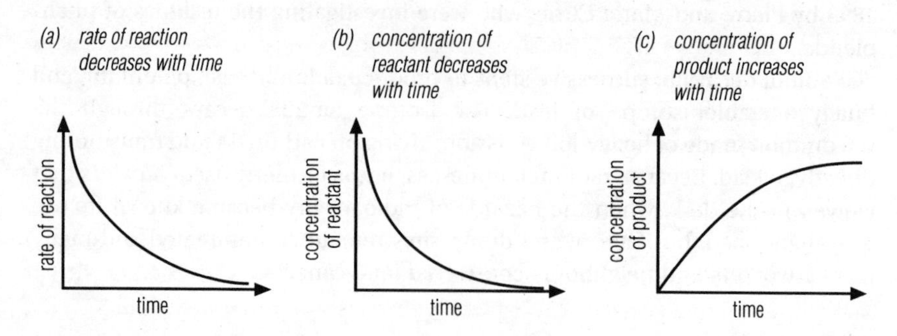

The rate of reaction decreases with time whilst the concentration of product increases.

Chemical equations show the reactants and products of a chemical reaction by using chemical symbols and formulae. State symbols and the energy symbol (ΔH) can be used to show whether reactants and products are solids, liquids, or gases, and whether energy has been released or absorbed during the reaction.

reactivity series
chemical series produced by arranging the metals in order of their ease of reaction with reagents such as oxygen, water, and acids. This arrangement aids the understanding of the properties of metals, helps to explain differences between them, and enables predictions to be made about a metal's behaviour, based on a knowledge of its position or properties. It also allows prediction of the relative stability of the compounds formed by an element: the more reactive the metal, the more stable its compounds are likely to be.

reduction
gain of electrons, loss of oxygen, or gain of hydrogen by an atom, ion, or molecule during a chemical reaction. Reduction may be brought about by reaction with another compound, which is simultaneously oxidized (reducing agent), or electrically at the cathode (negative electrode) of an electric cell. Examples include the reduction of iron(III) oxide to iron by carbon monoxide:

$$Fe_2O_3 + 3CO \rightarrow 2Fe + 3CO_2$$

the hydrogenation of ethene to ethane:

$$CH_2{=}CH_2 + H_2 \rightarrow CH_3{-}CH_3$$

and the reduction of a sodium ion to sodium.

$$Na^+ + e^- \rightarrow Na$$

relative atomic mass
mass of an atom relative to one-twelfth the mass of an atom of carbon-12. It depends primarily on the number of protons and neutrons in the atom, the

electrons having negligible mass. If more than one isotope of the element is present, the relative atomic mass is calculated by taking an average that takes account of the relative proportions of each isotope, resulting in values that are not whole numbers. The term ***atomic weight***, although commonly used, is strictly speaking incorrect.

relative molecular mass

mass of a molecule, calculated relative to one-twelfth the mass of an atom of carbon-12. It is found by adding the relative atomic masses of the atoms that make up the molecule. The term ***molecular weight*** is often used, but strictly this is incorrect.

reversible reaction

chemical reaction that proceeds in both directions at the same time, as the product decomposes back into reactants as it is being produced. Such reactions do not run to completion, provided that no substance leaves the system. Examples include the manufacture of ammonia from hydrogen and nitrogen, and the oxidation of sulphur dioxide to sulphur trioxide.

rhenium (Latin *Rhenus* 'Rhine')

heavy, silver-white, metallic element, symbol Re, atomic number 75, relative atomic mass 186.2. It has chemical properties similar to those of manganese and a very high melting point (3,180°C/5,756°F), which makes it valuable as an ingredient in alloys.

It was identified and named in 1925 by German chemists Walter Noddack, Ida Tacke, and Otto Berg from the Latin name for the River Rhine.

rhodium (Greek *rhodon* 'rose')

hard, silver-white, metallic element, symbol Rh, atomic number 45, relative atomic mass 102.905. It is one of the so-called platinum group of metals and is resistant to tarnish, corrosion, and acid. It occurs as a free metal in the natural alloy osmiridium and is used in jewellery, electroplating, and thermocouples.

rubidium (Latin *rubidus* 'red')

soft, silver-white, metallic element, symbol Rb, atomic number 37, relative atomic mass 85.47. It is one of the alkali metals, ignites spontaneously in air, and reacts violently with water. It is used in photocells and vacuum-tube filaments. Rubidium was discovered spectroscopically by German physicists Robert Bunsen and Gustav Kirchhoff in 1861 and named after the red lines in its spectrum.

rust

reddish-brown oxide of iron formed by the action of moisture and oxygen on the metal. It consists mainly of hydrated iron(III) oxide ($Fe_2O_3.H_2O$) and iron(III) hydroxide ($Fe(OH)_3$). Rusting is the commonest form of corrosion.

ruthenium

hard, brittle, silver-white, metallic element, symbol Ru, atomic number 44, relative atomic mass 101.07. It is one of the so-called platinum group of metals; it occurs in platinum ores as a free metal and in the natural alloy osmiridium. It is used as a hardener in alloys and as a catalyst; its compounds are used as colouring agents in glass and ceramics.

rutherfordium

synthesized, radioactive, metallic element, symbol Rf. It is the first of the transactinide series, atomic number 104, relative atomic mass 262. It is produced by bombarding californium with carbon nuclei and has ten isotopes, the longest-lived of which, Rf-262, has a half-life of 70 seconds. Two institutions claim to be the first to have synthesized it: the Joint Institute for Nuclear Research in Dubna, Russia, in 1964; and the University of California at Berkeley, USA, in 1969.

salt

any compound formed from an acid and a base through the replacement of all or part of the hydrogen in the acid by a metal or electropositive radical. ***Common salt*** is sodium chloride (see salt, common).

A salt may be produced by chemical reaction between an acid and a base, or by the displacement of hydrogen from an acid by a metal (see displacement reaction). As a solid, the ions normally adopt a regular arrangement to form crystals. Some salts form stable crystals only as hydrates (when combined with water). Most inorganic salts readily dissolve in water to give an electrolyte (a solution that conducts electricity).

salt, common or *sodium chloride*

NaCl white crystalline solid, found dissolved in sea water and as rock salt (the mineral halite) in large deposits and salt domes. Common salt is used extensively in the food industry as a preservative and for flavouring, and in the chemical industry in the making of chlorine and sodium.

samarium

hard, brittle, grey-white, metallic element of the lanthanide series, symbol Sm, atomic number 62, relative atomic mass 150.4. It is widely distributed in nature and is obtained commercially from the minerals monzanite and bastnaesite. It is used only occasionally in industry, mainly as a catalyst in organic reactions. Samarium was discovered by spectroscopic analysis of the mineral samarskite and named 1879 by French chemist Paul Lecoq de Boisbaudran (1838–1912) after its source.

saponification

hydrolysis (splitting) of an ester by treatment with a strong alkali, resulting in the liberation of the alcohol from which the ester had been derived and a salt of the constituent fatty acid. The process is used in the manufacture of soap.

saturated fatty acid
fatty acid in which there are no double bonds in the hydrocarbon chain.

saturated solution
solution obtained when a solvent (liquid) can dissolve no more of a solute (usually a solid) at a particular temperature. Normally, a slight fall in temperature causes some of the solute to crystallize out of solution. If this does not happen the phenomenon is called supercooling, and the solution is said to be *supersaturated*.

scale
calcium carbonate deposits that form on the inside of a kettle or boiler as a result of boiling hard water.

scandium
silver-white, metallic element of the lanthanide series, symbol Sc, atomic number 21, relative atomic mass 44.956. Its compounds are found widely distributed in nature, but only in minute amounts. The metal has little industrial importance.

Scandium oxide (scandia) is used as a catalyst, in making crucibles and other ceramic parts, and scandium sulphate (in very dilute aqueous solution) is used in agriculture to improve seed germination. Scandium is relatively more abundant in the Sun and other stars than on Earth.

seaborgium
synthesized radioactive element of the transactinide series, symbol Sg, atomic number 106, relative atomic mass 263. It was first synthesized in 1974 in the USA. The discovery was not confirmed until 1993. It was officially named in 1997 after US nuclear chemist Glenn Seaborg.

The University of California, Berkeley, bombarded californium with oxygen nuclei to get isotope 263; the Joint Institute for Nuclear Research, Dubna, Russia, bombarded lead with chromium nuclei to obtain isotopes 259 and 260.

selenium (Greek *Selene* 'Moon')
grey, nonmetallic element, symbol Se, atomic number 34, relative atomic mass 78.96. It belongs to the sulphur group and occurs in several allotropic forms that differ in their physical and chemical properties. It is an essential trace element in human nutrition.

Obtained from many sulphide ores and selenides, it is used as a red colouring for glass and enamel. Because its electrical conductivity varies with the intensity of light, selenium is used extensively in photoelectric devices. It was discovered in 1817 by Swedish chemist Jöns Berzelius and named after the Moon because its properties follow those of tellurium, whose name derives from Latin *Tellus* 'Earth'.

semiconductor

material with electrical conductivity intermediate between metals and insulators and used in a wide range of electronic devices. Certain crystalline materials, most notably silicon and germanium, have a small number of free electrons that have escaped from the bonds between the atoms. The atoms from which they have escaped possess vacancies, called holes, which are similarly able to move from atom to atom and can be regarded as positive charges. Current can be carried by both electrons (negative carriers) and holes (positive carriers). Such materials are known as ***intrinsic semiconductors.***

silicon (Latin *silex* 'flint')

brittle, nonmetallic element, symbol Si, atomic number 14, relative atomic mass 28.086. It is the second-most abundant element (after oxygen) in the Earth's crust and occurs in amorphous and crystalline forms. In nature it is found only in combination with other elements, chiefly with oxygen in silica (silicon dioxide, SiO_2) and the silicates. These form the mineral quartz, which makes up most sands, gravels, and beaches.

The element was isolated by Swedish chemist Jöns Berzelius in 1823, having been named in 1817 by Scottish chemist Thomas Thomson by analogy with boron and carbon because of its chemical resemblance to these elements.

silver

white, lustrous, extremely malleable and ductile, metallic element, symbol Ag (from Latin *argentum*), atomic number 47, relative atomic mass 107.868. It occurs in nature in ores and as a free metal; the chief ores are sulphides, from which the metal is extracted by smelting with lead. It is the best metallic conductor of both heat and electricity; its most useful compounds are the chloride and bromide, which darken on exposure to light and are the basis of photographic emulsions.

Silver is used ornamentally, for jewellery and tableware, for coinage, in electroplating, electrical contacts, and dentistry, and as a solder. It has been mined since prehistory; its name is an ancient non-Indo-European one, *silubr*, borrowed by the Germanic branch as *silber.*

smelting

processing a metallic ore in a furnace to produce the metal. Oxide ores such as iron ore are smelted with coke (carbon), which reduces the ore into metal and also provides fuel for the process. A substance such as limestone is often added during smelting to facilitate the melting process and to form a slag, which dissolves many of the impurities present.

soap

mixture of the sodium salts of various fatty acids: palmitic, stearic, and oleic acid. It is made by the action of sodium hydroxide (caustic soda) or potassium hydroxide (caustic potash) on fats of animal or vegetable origin. Soap makes grease and dirt disperse in water in a similar manner to a detergent.

sodium
soft, waxlike, silver-white, metallic element, symbol Na (from Latin *natrium*), atomic number 11, relative atomic mass 22.989. It is one of the alkali metals and has a very low density, being light enough to float on water. It is the sixth-most abundant element (the fourth-most abundant metal) in the Earth's crust. Sodium is highly reactive, oxidizing rapidly when exposed to air and reacting violently with water. Its most familiar compound is sodium chloride (common salt), which occurs naturally in the oceans and in salt deposits left by dried-up ancient seas.

sodium chloride or *common salt* or *table salt*
$NaCl$ white, crystalline compound found widely in nature. It is a typical ionic solid with a high melting point (801°C/1,474°F); it is soluble in water, insoluble in organic solvents, and is a strong electrolyte when molten or in aqueous solution. Found in concentrated deposits as the mineral halite, it is widely used in the food industry as a flavouring and preservative, and in the chemical industry in the manufacture of sodium, chlorine, and sodium carbonate.

sodium hydroxide or *caustic soda*
$NaOH$ the commonest alkali. The solid and the solution are corrosive. It is used to neutralize acids, in the manufacture of soap, and in drain and oven cleaners. It is prepared industrially from sodium chloride by the electrolysis of concentrated brine.

solid
state of matter that holds its own shape (as opposed to a liquid, which takes up the shape of its container, or a gas, which totally fills its container). According to kinetic theory, the atoms or molecules in a solid are not free to move but merely vibrate about fixed positions, such as those in crystal lattices.

solidification
change of state from liquid (or vapour) to solid that occurs at the freezing point of a substance.

solubility
measure of the amount of solute (usually a solid or gas) that will dissolve in a given amount of solvent (usually a liquid) at a particular temperature. Solubility may be expressed as grams of solute per 100 grams of solvent or, for a gas, in parts per million (ppm) of solvent.

solute
substance that is dissolved in another substance (see solution).

solution
two or more substances mixed to form a single, homogenous phase. One of the substances is the ***solvent*** and the others (***solutes***) are said to be dissolved in it.

The constituents of a solution may be solid, liquid, or gaseous. The solvent is normally the substance that is present in greatest quantity; however, if one of the constituents is a liquid this is considered to be the solvent even if it is not the major substance.

Solvay process

industrial process for the manufacture of sodium carbonate. It is a multistage process in which carbon dioxide is generated from limestone and passed through brine saturated with ammonia. Sodium hydrogen carbonate is isolated and heated to yield sodium carbonate. All intermediate by-products are recycled so that the only ultimate by-product is calcium chloride.

solvent

substance, usually a liquid, that will dissolve another substance (see solution). Although the commonest solvent is water, in popular use the term refers to low-boiling-point organic liquids, which are harmful if used in a confined space. They can give rise to respiratory problems, liver damage, and neurological complaints.

Typical organic solvents are petroleum distillates (in glues), xylol (in paints), alcohols (for synthetic and natural resins such as shellac), esters (in lacquers, including nail varnish), ketones (in cellulose lacquers and resins), and chlorinated hydrocarbons (as paint stripper and dry-cleaning fluids). The fumes of some solvents, when inhaled (glue-sniffing), affect mood and perception. In addition to damaging the brain and lungs, repeated inhalation of solvent from a plastic bag can cause death by asphyxia.

spectrometry

technique involving the measurement of the spectrum of energies (not necessarily electromagnetic radiation) emitted or absorbed by a substance.

spectroscopy

study of spectra associated with atoms or molecules in solid, liquid, or gaseous phase. Spectroscopy can be used to identify unknown compounds and is an invaluable tool in science, medicine, and industry (for example, in checking the purity of drugs).

Emission spectroscopy is the study of the characteristic series of sharp lines in the spectrum produced when an element is heated. Thus an unknown mixture can be analysed for its component elements. Related is ***absorption spectroscopy***, dealing with atoms and molecules as they absorb energy in a characteristic way. Again, dark lines can be used for analysis. More detailed structural information can be obtained using ***infrared spectroscopy*** (concerned with molecular vibrations) or ***nuclear magnetic resonance (NMR) spectroscopy*** (concerned with interactions between adjacent atomic nuclei). ***Supersonic jet laser beam spectroscopy*** enables the isolation and study of clusters in the gas phase. A laser vaporizes a small sample, which is cooled in helium, and ejected into an evacuated chamber. The jet of clusters expands supersonically, cooling the clusters to near absolute zero, and stabilizing them for study in a mass spectrometer.

stain
coloured compound that will bind to other substances. Stains are used extensively in microbiology to colour micro-organisms and in histochemistry to detect the presence and whereabouts in plant and animal tissue of substances such as fats, cellulose, and proteins.

standard temperature and pressure *STP*
standard set of conditions for experimental measurements, to enable comparisons to be made between sets of results. Standard temperature is 0°C/32°F (273 K) and standard pressure 1 atmosphere (101,325 Pa).

starch
widely distributed, high-molecular-mass carbohydrate, produced by plants as a food store; main dietary sources are cereals, legumes, and tubers, including potatoes. It consists of varying proportions of two glucose polymers (polysaccharides): straight-chain (amylose) and branched (amylopectin) molecules. The usual chemical test for starch consists of adding a drop of iodine solution, which turns bright blue.

steam
dry, invisible gas formed by vaporizing water. The visible cloud that normally forms in the air when water is vaporized is due to minute suspended water particles. Steam is widely used in chemical and other industrial processes and for the generation of power.

stearin
mixture of stearic and palmitic acids, used to make soap.

steel
alloy or mixture of iron and up to 1.7% carbon, sometimes with other elements, such as manganese, phosphorus, sulphur, and silicon. The USA, Russia, Ukraine, and Japan are the main steel producers. Steel has innumerable uses, including ship and car manufacture, skyscraper frames, and machinery of all kinds.

Steels with only small amounts of other metals are called ***carbon steels***. These steels are far stronger than pure iron, with properties varying with the composition. ***Alloy steels*** contain greater amounts of other metals. Low-alloy steels have less than 5% of the alloying material; high-alloy steels have more. Low-alloy steels containing up to 5% silicon with relatively little carbon have a high electrical resistance and are used in power transformers and motor or generator cores, for example. ***Stainless steel*** is a high-alloy steel containing at least 11% chromium. Steels with up to 20% tungsten are very hard and are used in high-speed cutting tools. About 50% of the world's steel is now made from scrap.

Steel is produced by removing impurities, such as carbon, from raw or pig iron, produced by a blast furnace. The main industrial process is the ***basic–oxygen process***, in which molten pig iron and scrap steel is placed in a container lined with heat-resistant, alkaline (basic) bricks. A pipe or lance is lowered near to the surface of the molten metal and pure oxygen blown through it at high pressure. The surface of the metal is disturbed by the blast and the impurities are oxidized (burned out).

strontium

soft, ductile, pale-yellow, metallic element, symbol Sr, atomic number 38, relative atomic mass 87.62. It is one of the alkaline-earth metals, widely distributed in small quantities only as a sulphate or carbonate. Strontium salts burn with a red flame and are used in fireworks and signal flares.

The radioactive isotopes Sr-89 and Sr-90 (half-life 25 years) are some of the most dangerous products of the nuclear industry; they are fission products in nuclear explosions and in the reactors of nuclear power plants. Strontium is chemically similar to calcium and deposits in bones and other tissues, where the radioactivity is damaging. The element was named in 1808 by English chemist Humphry Davy, who isolated it by electrolysis, after Strontian, a mining location in Scotland where it was first found.

sublimation

conversion of a solid to vapour without passing through the liquid phase. Sublimation depends on the fact that the boiling-point of the solid substance is lower than its melting-point at atmospheric pressure. Thus by increasing pressure, a substance that sublimes can be made to go through a liquid stage before passing into the vapour state.

Some substances that do not sublime at atmospheric pressure can be made to do so at low pressures. This is the principle of freeze-drying, during which ice sublimes at low pressure.

substitution reaction

replacement of one atom or functional group in an organic molecule by another.

substrate

compound or mixture of compounds acted on by an enzyme. The term also refers to a substance such as agar that provides the nutrients for the metabolism of micro-organisms. Since the enzyme systems of micro-organisms regulate their metabolism, the essential meaning is the same.

sucrose or *cane sugar* or *beet sugar*
$C_{12}H_{22}O_{11}$ a sugar found in the pith of sugar cane and in sugar beet. It is popularly known as sugar. Sucrose is a disaccharide sugar, each of its molecules being made up of two simple sugar (monosaccharide) units: glucose and fructose.

sulphate
SO_4^{2-} salt or ester derived from sulphuric acid. Most sulphates are water soluble (the chief exceptions are lead, calcium, strontium, and barium sulphates) and require a very high temperature to decompose them.

The commonest sulphates seen in the laboratory are copper(II) sulphate ($CuSO_4$), iron(II) sulphate ($FeSO_4$), and aluminium sulphate ($Al_2(SO_4)_3$). The ion is detected in solution by using barium chloride or barium nitrate to precipitate the insoluble sulphate.

sulphide
compound of sulphur and another element in which sulphur is the more electronegative element (see electronegativity). Sulphides occur in a number of minerals. Some of the more volatile sulphides have extremely unpleasant odours (hydrogen sulphide smells of bad eggs).

sulphite
SO_3^{2-} salt or ester derived from sulphurous acid.

sulphonamide
any of a group of compounds containing the chemical group sulphonamide (SO_2NH_2) or its derivatives, which were, and still are in some cases, used to treat bacterial diseases. Sulphadiazine ($C_{10}H_{10}N_4O_2S$) is an example.

Sulphonamide was the first commercially available antibacterial drug, the forerunner of a range of similar drugs. Toxicity and increasing resistance have limited their use chiefly to the treatment of urinary-tract infection.

sulphur
brittle, pale-yellow, nonmetallic element, symbol S, atomic number 16, relative atomic mass 32.064. It occurs in three allotropic forms: two crystalline (called rhombic and monoclinic, following the arrangements of the atoms within the crystals) and one amorphous. It

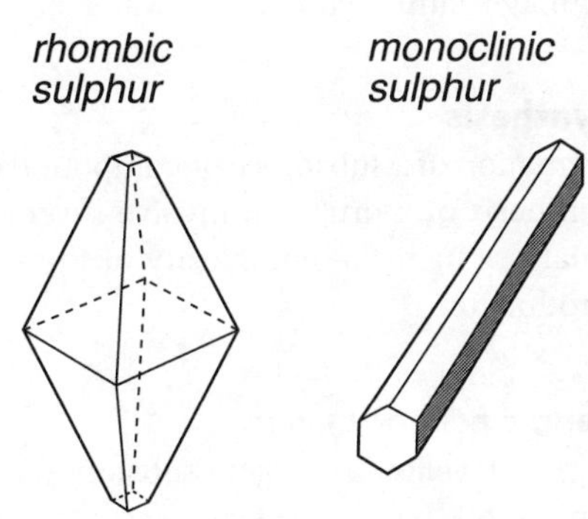

Two common allotropes of sulphur: rhombic and monoclinic crystals. A reactive element, sulphur combines with most other elements and has a wide range of industrial uses. It often occurs around hot springs and in volcanic regions, and there are large deposits in the USA (Texas and Louisiana), Japan, Sicily, and Mexico.

burns in air with a blue flame and a stifling odour. Insoluble in water but soluble in carbon disulphide, it is a good electrical insulator. Sulphur is widely used in the manufacture of sulphuric acid (used to treat phosphate rock to make fertilizers) and in making paper, matches, gunpowder and fireworks, in vulcanizing rubber, and in medicines and insecticides.

sulphur dioxide

SO_2 pungent gas produced by burning sulphur or sulphide ores in air or oxygen. It is widely used for making sulphuric acid and for disinfecting food vessels and equipment, and as a preservative in some food products. It occurs in industrial flue gases and is a major cause of acid rain.

sulphuric acid or *oil of vitriol*

H_2SO_4 a dense, viscous, colourless liquid that is extremely corrosive. It gives out heat when added to water and can cause severe burns. Sulphuric acid is used extensively in the chemical industry, in the refining of petrol, and in the manufacture of fertilizers, detergents, explosives, and dyes. It forms the acid component of car batteries.

sulphurous acid

H_2SO_3 solution of sulphur dioxide (SO_2) in water. It is a weak acid.

suspension

mixture consisting of small solid particles dispersed in a liquid or gas, which will settle on standing. An example is milk of magnesia, which is a suspension of magnesium hydroxide in water.

synthesis

formation of a substance or compound from more elementary compounds. The synthesis of a drug can involve several stages from the initial material to the final product; the complexity of these stages is a major factor in the cost of production.

tannic acid or *tannin*

$C_{14}H_{10}O_9$ yellow astringent substance, composed of several phenol rings, occurring in the bark, wood, roots, fruits, and galls (growths) of certain trees, such as the oak. It precipitates gelatin to give an insoluble compound used in the manufacture of leather from hides (tanning).

tantalum

hard, ductile, lustrous, grey-white, metallic element, symbol Ta, atomic number 73, relative atomic mass 180.948. It occurs with niobium in tantalite and other minerals. It can be drawn into wire with a very high melting point and great tenacity, useful for lamp filaments subject to vibration. It is also used in alloys, for corrosion-resistant laboratory apparatus and chemical equipment, as a catalyst in manufacturing synthetic rubber, in tools and instruments, and in rectifiers and capacitors.

It was discovered and named in 1802 by Swedish chemist Anders Ekeberg (1767–1813) after the mythological Greek character Tantalos.

tar
dark brown or black viscous liquid obtained by the destructive distillation of coal, shale, and wood. Tars consist of a mixture of hydrocarbons, acids, and bases. Creosote and paraffin oil are produced from wood tar. See also coal tar.

tautomerism
form of isomerism in which two interconvertible isomers are in equilibrium. It is often specifically applied to an equilibrium between the keto ($-CH_2-C=O$) and enol ($-CH=C-OH$) forms of carbonyl compounds.

technetium (Greek *technetos* 'artificial')
silver-grey, radioactive, metallic element, symbol Tc, atomic number 43, relative atomic mass 98.906. It occurs in nature only in extremely minute amounts, produced as a fission product from uranium in pitchblende and other uranium ores. Its longest-lived isotope, Tc-99, has a half-life of 216,000 years. It is a superconductor and is used as a hardener in steel alloys and as a medical tracer.

It was synthesized in 1937 (named in 1947) by Italian physicists Carlo Perrier and Emilio Segrè, who bombarded molybdenum with deuterons, looking to fill a missing spot in the periodic table of the elements (at that time it was considered not to occur in nature). It was later isolated in large amounts from the fission product debris of uranium fuel in nuclear reactors.

Teflon
trade name for polytetrafluoroethene (PTFE), a tough, waxlike, heat-resistant plastic used for coating nonstick cookware and in gaskets and bearings.

tellurium (Latin *Tellus* 'Earth')
silver-white, semi-metallic (metalloid) element, symbol Te, atomic number 52, relative atomic mass 127.60. Chemically it is similar to sulphur and selenium, and it is considered one of the sulphur group. It occurs naturally in telluride minerals, and is used in colouring glass blue-brown, in the electrolytic refining of zinc, in electronics, and as a catalyst in refining petroleum.

It was discovered in 1782 by the Austrian mineralogist Franz Müller (1740–1825), and named in 1798 by the German chemist Martin Klaproth.

terbium
soft, silver-grey, metallic element of the lanthanide series, symbol Tb, atomic number 65, relative atomic mass 158.925. It occurs in gadolinite and other ores, with yttrium and ytterbium, and is used in lasers, semiconductors, and televi-

sion tubes. It was named in 1843 by Swedish chemist Carl Mosander (1797–1858) after the town of Ytterby, Sweden, where it was first found.

tetrachloromethane

CCl_4 or ***carbon tetrachloride*** chlorinated organic compound that is a very efficient solvent for fats and greases, and was at one time the main constituent of household dry-cleaning fluids and of fire extinguishers used with electrical and petrol fires. Its use became restricted after it was discovered to be carcinogenic and it has now been largely removed from educational and industrial laboratories.

thallium (Greek *thallos* 'young green shoot')

soft, bluish-white, malleable, metallic element, symbol Tl, atomic number 81, relative atomic mass 204.38. It is a poor conductor of electricity. Its compounds are poisonous and are used as insecticides and rodent poisons; some are used in the optical-glass and infrared-glass industries and in photocells.

Discovered spectroscopically by its green line, thallium was isolated and named by William Crookes in 1861.

thorium

dark-grey, radioactive, metallic element of the actinide series, symbol Th, atomic number 90, relative atomic mass 232.038. It occurs throughout the world in small quantities in minerals such as thorite and is widely distributed in monazite beach sands. It is one of three fissile elements (the others are uranium and plutonium) and its longest-lived isotope has a half-life of 1.39×10^{10} years. Thorium is used to strengthen alloys. It was discovered by Jöns Berzelius in 1828 and was named by him after the Norse god Thor.

thulium

soft, silver-white, malleable and ductile, metallic element of the lanthanide series, symbol Tm, atomic number 69, relative atomic mass 168.94. It is the least abundant of the rare earth metals, and was first found in gadolinite and various other minerals. It is used in arc lighting.

The X-ray-emitting isotope Tm-170 is used in portable X-ray units. Thulium was named by French chemist Paul Lecoq de Boisbaudran in 1886 after the northland, Thule.

tin

soft, silver-white, malleable and somewhat ductile, metallic element, symbol Sn (from Latin *stannum*), atomic number 50, relative atomic mass 118.69. Tin exhibits allotropy, having three forms: the familiar lustrous metallic form above 13.2°C/55.8°F; a brittle form above 161°C/321.8°F; and a grey powder form below 13.2°C/55.8°F (commonly called tin pest or tin disease). The metal is quite soft (slightly harder than lead) and can be rolled, pressed, or hammered into extremely thin sheets; it has a low melting point. In nature it occurs rarely as a free metal. It resists corrosion and is therefore used for coating and plating other metals.

Tin and copper smelted together form the oldest desired alloy, bronze; since the Bronze Age (3500 BC) that alloy has been the basis of both useful and decorative materials. Tin is also alloyed with metals other than copper to make solder and pewter. It was recognized as an element by Antoine Lavoisier, but the name is very old and comes from the Germanic form *zinn*. The mines of Cornwall were the principal Western source of tin until the 19th century, when rich deposits were found in South America, Africa, South-East Asia, and Australia. Tin production is concentrated in Malaysia, Indonesia, Brazil, and Bolivia.

titanium
strong, lightweight, silver-grey, metallic element, symbol Ti, atomic number 22, relative atomic mass 47.90. The ninth most abundant element in the Earth's crust, its compounds occur in practically all igneous rocks and their sedimentary deposits. It is very strong and resistant to corrosion, so it is used in building high-speed aircraft and spacecraft; it is also widely used in making alloys, as it unites with almost every metal except copper and aluminium. Titanium oxide is used in high-grade white pigments. The element was discovered in 1791 by English mineralogist William Gregor (1761–1817) and was named by German chemist Martin Klaproth in 1796 after the Titans, the giants of Greek mythology. It was not obtained in pure form until 1925.

titration
technique to find the concentration of one compound in a solution by determining how much of it will react with a known amount of another compound in solution. One of the solutions is measured by pipette into the reaction vessel. The other is added a little at a time from a burette. The end-point of the reaction is determined with an indicator or an electrochemical device.

tracer
small quantity of a radioactive isotope used to follow the path of a chemical reaction or a physical or biological process. The location (and possibly concentration) of the tracer is usually detected by using a Geiger–Muller counter. For example, the activity of the thyroid gland can be monitored by giving the patient an injection containing a small dose of a radioactive isotope of iodine, which is selectively absorbed from the bloodstream by the gland.

transactinide element
any of a series of eight radioactive, metallic elements with atomic numbers that extend beyond the actinide series, those from 104 (rutherfordium) upwards. They are grouped because of their expected chemical similarities (they are all bivalent), the properties differing only slightly with atomic number. All have half-lives that measure less than two minutes.

transuranic element or *transuranium element*
chemical element with an atomic number of 93 or more – that is, with a greater number of protons in the nucleus than has uranium. All transuranic elements

are radioactive. Neptunium and plutonium are found in nature; the others are synthesized in nuclear reactions.

tritium

radioactive isotope of hydrogen, three times as heavy as ordinary hydrogen, consisting of one proton and two neutrons. It has a half-life of 12.5 years.

tungsten (Swedish *tung sten* 'heavy stone')

hard, heavy, grey-white, metallic element, symbol W (from German *Wolfram*), atomic number 74, relative atomic mass 183.85. It occurs in the minerals wolframite, scheelite, and hubertite. It has the highest melting point of any metal (3,410°C/6,170°F) and is added to steel to make it harder, stronger, and more elastic; its other uses include high-speed cutting tools, electrical elements, and thermionic couplings. Its salts are used in the paint and tanning industries.

Tungsten was first recognized in 1781 by Swedish chemist Karl Scheele in the ore scheelite. It was isolated in 1783 by Spanish chemists Fausto D'Elhuyar (1755–1833) and his brother Juan José (1754–1796).

unsaturated compound

chemical compound in which two adjacent atoms are linked by a double or triple covalent bond. Examples are alkenes and alkynes, where the two adjacent atoms are both carbon, and ketones, where the unsaturation exists between atoms of different elements (carbon and oxygen). The laboratory test for unsaturated compounds is the addition of bromine water; if the test substance is unsaturated, the bromine water will be decolorized.

uranium

hard, lustrous, silver-white, malleable and ductile, radioactive, metallic element of the actinide series, symbol U, atomic number 92, relative atomic mass 238.029. It is the most abundant radioactive element in the Earth's crust, its decay giving rise to essentially all radioactive elements in nature; its final decay product is the stable element lead. Uranium combines readily with most elements to form compounds that are extremely poisonous. The chief ore is pitchblende, in which the element was discovered by German chemist Martin Klaproth in 1789; he named it after the planet Uranus, which had been discovered in 1781.

Uranium is one of three fissile elements (the others are thorium and plutonium). It was long considered to be the element with the highest atomic number to occur in nature on Earth. The isotopes U-238 and U-235 have been used to help determine the age of the Earth.

Uranium-238, which comprises about 99% of all naturally occurring uranium, has a half-life of 4.51×10^9 years. Because of its abundance, it is the isotope from which fissile plutonium is produced in breeder nuclear reactors. The fissile isotope U-235 has a half-life of 7.13×10^8 years and comprises about

0.7% of naturally occurring uranium; it is used directly as a fuel for nuclear reactors and in the manufacture of nuclear weapons.

valence

measure of an element's ability to combine with other elements, expressed as the number of atoms of hydrogen (or any other standard univalent element) capable of uniting with (or replacing) its atoms. The number of electrons in the outermost shell of the atom dictates the combining ability of an element.

The elements are described as uni-, di-, tri-, and tetravalent when they unite with one, two, three, and four univalent atoms respectively. Some elements have ***variable valence***: for example, nitrogen and phosphorus have a valency of both three and five. The valency of oxygen is two: hence the formula for water, H_2O (hydrogen being univalent).

valence electron

electron in the outermost shell of an atom. It is the valence electrons that are involved in the formation of ionic and covalent bonds (see molecule). The number of electrons in this outermost shell represents the maximum possible valence for many elements and matches the number of the group that the element occupies in the periodic table of the elements.

valence shell

outermost shell of electrons in an atom. It contains the valence electrons. Elements with four or more electrons in their outermost shell can show variable valence. Chlorine can show valences of 1, 3, 5, and 7 in different compounds.

vanadium

silver-white, malleable and ductile, metallic element, symbol V, atomic number 23, relative atomic mass 50.942. It occurs in certain iron, lead, and uranium ores and is widely distributed in small quantities in igneous and sedimentary rocks. It is used to make steel alloys, to which it adds tensile strength.

Spanish mineralogist Andrés del Rio (1764–1849) and Swedish chemist Nils Sefström (1787–1845) discovered vanadium independently, the former in 1801 and the latter in 1831. Del Rio named it 'erythronium', but was persuaded by other chemists that he had not in fact discovered a new element; Sefström gave it its present name, after the Norse goddess of love and beauty, Vanadis (or Freya).

van der Waals' law

modified form of the gas laws that includes corrections for the non-ideal behaviour of real gases (the molecules of ideal gases occupy no space and exert no forces on each other). It is named after Dutch physicist J D van der Waals (1837–1923).

The equation derived from the law states that:

$$(P + \frac{a}{V^2})(V - b) = RT$$

where *P*, *V*, and *T* are the pressure, volume, and temperature (in kelvin) of the gas, respectively; *R* is the gas constant; and *a* and *b* are constants for that particular gas.

vitriol

any of a number of sulphate salts. Blue, green, and white vitriols are copper, ferrous, and zinc sulphate, respectively. ***Oil of vitriol*** is sulphuric acid.

volatile

term describing a substance that readily passes from the liquid to the vapour phase. Volatile substances have a high vapour pressure.

water

chemical compound of hydrogen and oxygen elements, H_2O. It can exist as a solid (ice), liquid (water), or gas (water vapour). Water is the most common element on Earth and vital to all living organisms. It covers 70% of the Earth's surface, and provides a habitat for large numbers of aquatic organisms. It is the largest constituent of all living organisms – the human body consists of about 65% water. Pure water is a colourless, odourless, tasteless liquid which freezes at 0°C/32°F, and boils at 100°C/212°F. Natural water in the environment is never pure and always contains a variety of dissolved substances.

water of crystallization

water chemically bonded to a salt in its crystalline state. For example, in copper(II) sulphate, there are five moles of water per mole of copper sulphate: hence its formula is $CuSO_4.5H_2O$. This water is responsible for the colour and shape of the crystalline form. When the crystals are heated gently, the water is driven off as steam and a white powder of the anhydrous salt is formed.

$$CuSO_4.5H_2O_{(s)} \rightarrow CuSO_{4\ (s)} + 5H_2O_{(g)}$$

wax

solid fatty substance of animal, vegetable, or mineral origin. Waxes are composed variously of esters, fatty acids, free alcohols, and solid hydrocarbons.

xenon (Greek *xenos* 'stranger')

colourless, odourless, gaseous, non-metallic element, symbol Xe, atomic number 54, relative atomic mass 131.30. It is grouped with the rare gases and was long believed not to enter into reactions, but is now known to form some compounds, mostly with fluorine. It is a heavy gas present in very small quantities in the air (about one part in 20 million).

Xenon is used in bubble chambers, light bulbs, vacuum tubes, and lasers. It was discovered in 1898 in a residue from liquid air by Scottish chemists William Ramsay and Morris Travers.

X-ray fluorescence spectrometry
technique used to determine the major and trace elements in the chemical composition of such materials as ceramics, obsidian, and glass. A sample is bombarded with X-rays, and the wavelengths of the released energy, or fluorescent X-rays, are detected and measured. Different elements have unique wavelengths, and their concentrations can be estimated from the intensity of the released X-rays. This analysis may, for example, help an archaeologist in identifying the source of the material.

ytterbium
soft, lustrous, silvery, malleable and ductile element of the lanthanide series, symbol Yb, atomic number 70, relative atomic mass 173.04. It occurs with (and resembles) yttrium in gadolinite and other minerals, and is used in making steel and other alloys.

In 1878 Swiss chemist Jean-Charles de Marignac gave the name ytterbium (after the Swedish town of Ytterby, near where it was found) to what he believed to be a new element. French chemist Georges Urbain (1872–1938) discovered in 1907 that this was in fact a mixture of two elements: ytterbium and lutetium.

yttrium
silver-grey, metallic element, symbol Y, atomic number 39, relative atomic mass 88.905. It is associated with and resembles the rare earth elements (lanthanides), occurring in gadolinite, xenotime, and other minerals. It is used in colour-television tubes and to reduce steel corrosion.

The name derives from the Swedish town of Ytterby, near where it was first discovered in 1788. Swedish chemist Carl Mosander (1797–1858) isolated the element in 1843.

zinc (Germanic *zint* 'point')
hard, brittle, bluish-white, metallic element, symbol Zn, atomic number 30, relative atomic mass 65.37. The principal ore is sphalerite or zinc blende (zinc sulphide, ZnS). Zinc is hardly affected by air or moisture at ordinary temperatures; its chief uses are in alloys such as brass and in coating metals (for example, galvanized iron) and making batteries. Its compounds include zinc oxide, used in ointments (as an astringent) and cosmetics, paints, glass, and printing ink.

Zinc is an essential trace element in most animals; adult humans have 2–3 g/0.07–0.1 oz zinc in their bodies. There are more than 300 known enzymes that contain zinc.

Zinc has been used as a component of brass since the Bronze Age, but it was not recognized as a separate metal until 1746, when it was described by German chemist Andreas Sigismund Marggraf (1709–1782). The name derives from the shape of the crystals on smelting.

zirconium (Germanic *zircon*, from Persian *zargun* 'golden')
lustrous, greyish-white, strong, ductile, metallic element, symbol Zr, atomic number 40, relative atomic mass 91.22. It occurs in nature as the mineral zircon

(zirconium silicate), from which it is obtained commercially. It is used in some ceramics, alloys for wire and filaments, steel manufacture, and nuclear reactors, where its low neutron absorption is advantageous.

It was isolated in 1824 by Swedish chemist Jöns Berzelius. The name was proposed by English chemist Humphry Davy in 1808.

Appendix

The Chemical Elements

An element is a substance that cannot be split chemically into simpler substances. The atoms of a particular element all have the same number of protons in their nuclei (their atomic, or proton, number).

(– = not applicable.)

Name	Symbol	Atomic number	Atomic mass (amu)[1]	Relative density[2]	Melting or fusing point (°C)
Actinium	Ac	89	227[3]	–	–
Aluminium	Al	13	26.9815	2.58	658
Americium	Am	95	243[3]	–	–
Antimony	Sb	51	121.75	6.62	629
Argon	Ar	18	39.948	gas	–188
Arsenic	As	33	74.9216	5.73	volatile, 450
Astatine	At	85	210[3]	–	–
Barium	Ba	56	137.34	3.75	850
Berkelium	Bk	97	249[3]	–	–
Beryllium	Be	4	9.0122	1.93	1,281
Bismuth	Bi	83	208.9806	9.80	268
Bohrium	Bh	107	262[3]	–	–
Boron	B	5	10.81	2.5	2,300
Bromine	Br	35	79.904	3.19	–7.3
Cadmium	Cd	48	112.40	8.64	320
Caesium	Cs	55	132.9055	1.88	26
Calcium	Ca	20	40.08	1.58	851
Californium	Cf	98	251[3]	–	–

Name	Symbol	Atomic number	Atomic mass (amu)[1]	Relative density[2]	Melting or fusing point (°C)
Carbon	C	6	12.011	3.52	infusible
Cerium	Ce	58	140.12	6.68	623
Chlorine	Cl	17	35.453	gas	–102
Chromium	Cr	24	51.996	6.5	1,510
Cobalt	Co	27	58.9332	8.6	1,490
Copper	Cu	29	63.546	8.9	1,083
Curium	Cm	96	247[3]	–	–
Dubnium	Db	105	262[3]	–	–
Dysprosium	Dy	66	162.50	–	–
Einsteinium	Es	99	254[3]	–	–
Erbium	Er	68	167.26	4.8	–
Europium	Eu	63	151.96	–	–
Fermium	Fm	100	253[3]	–	–
Fluorine	F	9	18.9984	gas	–223
Francium	Fr	87	223[3]	–	–
Gadolinium	Gd	64	157.25	–	–
Gallium	Ga	31	69.72	5.95	30
Germanium	Ge	32	72.59	5.47	958
Gold	Au	79	196.9665	19.3	1,062
Hafnium	Hf	72	178.49	12.1	2,500
Hassium	Hs	108	265[3]	–	–
Helium	He	2	4.0026	gas	–272
Holmium	Ho	67	164.9303	–	–
Hydrogen	H	1	1.0080	gas	–258
Indium	In	49	114.82	7.4	155
Iodine	I	53	126.9045	4.95	114

Name	Symbol	Atomic number	Atomic mass (amu)[1]	Relative density[2]	Melting or fusing point (°C)
Iridium	Ir	77	192.22	22.4	2,375
Iron	Fe	26	55.847	7.86	1,525
Krypton	Kr	36	83.80	gas	–169
Lanthanum	La	57	138.9055	6.1	810
Lawrencium	Lr	103	260[3]	–	–
Lead	Pb	82	207.2	11.37	327
Lithium	Li	3	6.941	0.585	186
Lutetium	Lu	71	174.97	–	–
Magnesium	Mg	12	24.305	1.74	651
Manganese	Mn	25	54.9380	7.39	1,220
Meitnerium	Mt	109	266[3]	–	–
Mendelevium	Md	101	256[3]	–	–
Mercury	Hg	80	200.59	13.596	–38.9
Molybdenum	Mo	42	95.94	10.2	2,500
Neodymium	Nd	60	144.24	6.96	840
Neon	Ne	10	20.179	gas	–248.6
Neptunium	Np	93	237[3]	–	–
Nickel	Ni	28	58.71	8.9	1,452
Niobium	Nb	41	92.9064	8.4	1,950
Nitrogen	N	7	14.0067	gas	–211
Nobelium	No	102	254[3]	–	–
Osmium	Os	76	190.2	22.48	2,700
Oxygen	O	8	15.9994	gas	–227
Palladium	Pd	46	106.4	11.4	1,549
Phosphorus	P	15	30.9738	1.8–2.3	44
Platinum	Pt	78	195.09	21.5	1,755

Name	Symbol	Atomic number	Atomic mass (amu)[1]	Relative density[2]	Melting or fusing point (°C)
Plutonium	Pu	94	242[3]	–	–
Polonium	Po	84	210[3]	–	–
Potassium	K	19	39.102	0.87	63
Praseodymium	Pr	59	140.9077	6.48	940
Promethium	Pm	61	145[3]	–	–
Protactinium	Pa	91	231.0359	–	–
Radium	Ra	88	226.0254	6.0	700
Radon	Rn	86	222[3]	gas	–150
Rhenium	Re	75	186.2	21	3,000
Rhodium	Rh	45	102.9055	12.1	1,950
Rubidium	Rb	37	85.4678	1.52	39
Ruthenium	Ru	44	101.07	12.26	2,400
Rutherfordium	Rf	104	262[3]	–	–
Samarium	Sm	62	150.4	7.7	1,350
Scandium	Sc	21	44.9559	–	–
Seaborgium	Sg	106	263[3]	–	–
Selenium	Se	34	78.96	4.5	170–220
Silicon	Si	14	28.086	2.0–2.4	1,370
Silver	Ag	47	107.868	10.5	960
Sodium	Na	11	22.9898	0.978	97
Strontium	Sr	38	87.62	2.54	800
Sulphur	S	16	32.06	2.07	115–119
Tantalum	Ta	73	180.9479	16.6	2,900
Technetium	Tc	43	99[3]	–	–
Tellurium	Te	52	127.60	6.0	446
Terbium	Tb	65	158.9254	–	–

Name	Symbol	Atomic number	Atomic mass (amu)[1]	Relative density[2]	Melting or fusing point (°C)
Thallium	Tl	81	204.37	11.85	302
Thorium	Th	90	232.0381	11.00	1,750
Thulium	Tm	69	168.9342	–	–
Tin	Sn	50	118.69	7.3	232
Titanium	Ti	22	47.90	4.54	1,850
Tungsten	W	74	183.85	19.1	2,900–3,000
Ununbium	Uub[4]	112	277[3]	–	–
Ununhexium	Uuh[4]	116	289[3]	–	–
Ununnilium	Uun[4]	110	269[3]	–	–
Ununoctium	Uno[4]	118	293[3]	–	–
Ununquadium	Unq[4]	114	289[3]	–	–
Unununium	Uuu[4]	111	272[3]	–	–
Uranium	U	92	238.029	18.7	–
Vanadium	V	23	50.9414	5.5	1,710
Xenon	Xe	54	131.30	gas	–140
Ytterbium	Yb	70	173.04	–	–
Yttrium	Y	39	88.9059	3.8	–
Zinc	Zn	30	65.37	7.12	418
Zirconium	Zr	40	91.22	4.15	2,130

[1] Atomic mass units.
[2] Also known as specific gravity. Relative density is the density (at 20°C/68°F) of a solid or liquid relative to the maximum density of water (at 4°C/39.2°F).
[3] The number given is that for the most stable isotope of the element.
[4] Elements as yet unnamed; temporary identification assigned until a name is approved by the International Union for Pure and Applied Chemistry.

Nobel Prize for Chemistry

Year	Winner(s)[1]	Awarded for
1901	Jacobus van't Hoff (Netherlands)	laws of chemical dynamics and osmotic pressure
1902	Emil Fischer (Germany)	sugar and purine syntheses
1903	Svante Arrhenius (Sweden)	theory of electrolytic dissociation
1904	William Ramsay (UK)	discovery of rare gases in air and their locations in the periodic table
1905	Adolf von Baeyer (Germany)	work in organic dyes and hydroaromatic compounds
1906	Henri Moissan (France)	isolation of fluorine and adoption of electric furnace
1907	Eduard Buchner (Germany)	biochemical research and discovery of cell-free fermentation
1908	Ernest Rutherford (UK)	work in atomic disintegration and the chemistry of radioactive substances
1909	Wilhelm Ostwald (Germany)	work in catalysis and principles of equilibria and rates of reaction
1910	Otto Wallach (Germany)	work in alicyclic compounds
1911	Marie Curie (France)	discovery of radium and polonium, and the isolation and study of radium
1912	Victor Grignard (France)	discovery of Grignard reagents
	Paul Sabatier (France)	finding method of catalytic hydrogenation of organic compounds
1913	Alfred Werner (Switzerland)	work in bonding of atoms within inorganic molecules
1914	Theodore Richards (USA)	accurate determination of the atomic masses of many elements
1915	Richard Willstätter (Germany)	research into plant pigments, especially chlorophyll
1916	no award	
1917	no award	

Year	Winner(s)[1]	Awarded for
1918	Fritz Haber (Germany)	synthesis of ammonia from its elements
1919	no award	
1920	Walther Nernst (Germany)	work in thermochemistry
1921	Frederick Soddy (UK)	work in radioactive substances, especially isotopes
1922	Francis Aston (UK)	work in mass spectrometry of isotopes of radioactive elements, and enunciation of the whole-number rule
1923	Fritz Pregl (Austria)	method of microanalysis of organic substances
1924	no award	
1925	Richard Zsigmondy (Austria)	elucidation of heterogeneity of colloids
1926	Theodor Svedberg (Sweden)	investigation of dispersed systems
1927	Heinrich Wieland (Germany)	research on constitution of bile acids and related substances
1928	Adolf Windaus (Germany)	research on constitution of sterols and related vitamins
1929	Arthur Harden (UK) and Hans von Euler-Chelpin (Sweden)	work on fermentation of sugar and fermentative enzymes
1930	Hans Fischer (Germany)	analysis of haem (the iron-bearing group in haemoglobin) and chlorophyll, and the synthesis of haemin (a compound of haem)
1931	Carl Bosch (Germany) and Friedrich Bergius (Germany)	invention and development of chemical high-pressure methods
1932	Irving Langmuir (USA)	discoveries and investigations in surface chemistry
1933	no award	

Year	Winner(s)[1]	Awarded for
1934	Harold Urey (USA)	discovery of deuterium (heavy hydrogen)
1935	Irène and Frédéric Joliot-Curie (France)	synthesis of new radioactive elements
1936	Peter Debye (Netherlands)	work in molecular structures by investigation of dipole moments and the diffraction of X-rays and electrons in gases
1937	Norman Haworth (UK)	work in carbohydrates and ascorbic acid (vitamin C)
	Paul Karrer (Switzerland)	work in carotenoids, flavins, retinol (vitamin A) and riboflavin (vitamin B_2)
1938	Richard Kuhn (Germany) (declined)	carotenoids and vitamins research
1939	Adolf Butenandt (Germany) (declined)	work in sex hormones
	Leopold Ružička (Switzerland)	polymethylenes and higher terpenes
1940	no award	
1941	no award	
1942	no award	
1943	Georg von Hevesy (Hungary)	use of isotopes as tracers in chemical processes
1944	Otto Hahn (Germany)	discovery of nuclear fission
1945	Artturi Virtanen (Finland)	work in agriculture and nutrition, especially fodder preservation
1946	James Sumner (USA)	discovery of crystallization of enzymes
	John Northrop (USA) and Wendell Stanley (USA)	preparation of pure enzymes and virus proteins
1947	Robert Robinson (UK)	investigation of biologically important plant products, especially alkaloids
1948	Arne Tiselius (Sweden)	researches in electrophoresis and adsorption analysis, and discoveries concerning serum proteins

Year	Winner(s)[1]	Awarded for
1949	William Giauque (USA)	work in chemical thermodynamics, especially at very low temperatures
1950	Otto Diels (West Germany) and Kurt Alder (West Germany)	discovery and development of diene synthesis
1951	Edwin McMillan (USA) and Glenn Seaborg (USA)	discovery and work in chemistry of transuranic elements
1952	Archer Martin (UK) and Richard Synge (UK)	development of partition chromatography
1953	Hermann Staudinger (West Germany)	discoveries in macromolecular chemistry
1954	Linus Pauling (USA)	study of nature of chemical bonds, especially in complex substances
1955	Vincent Du Vigneaud (USA)	investigations into biochemically important sulphur compounds, and the first synthesis of a polypeptide hormone
1956	Cyril Hinshelwood (UK) and Nikolai Semenov (USSR)	work in mechanism of chemical reactions
1957	Alexander Todd (UK)	work in nucleotides and nucleotide coenzymes
1958	Frederick Sanger (UK)	determination of the structure of proteins, especially insulin
1959	Jaroslav Heyrovský (Czechoslovakia)	discovery and development of polarographic methods of chemical analysis
1960	Willard Libby (USA)	development of radiocarbon dating in archaeology, geology, and geography
1961	Melvin Calvin (USA)	study of assimilation of carbon dioxide by plants
1962	Max Perutz (UK) and John Kendrew (UK)	determination of structures of globular proteins
1963	Karl Ziegler (West Germany) and Giulio Natta (Italy)	chemistry and technology of producing high polymers

Year	Winner(s)[1]	Awarded for
1964	Dorothy Crowfoot Hodgkin (UK)	crystallographic determination of the structures of biochemical compounds, notably penicillin and cyanocobalamin (vitamin B_{12})
1965	Robert Woodward (USA)	organic synthesis
1966	Robert Mulliken (USA)	molecular orbital theory of chemical bonds and structures
1967	Manfred Eigen (West Germany), Ronald Norrish (UK), and George Porter (UK)	investigation of rapid chemical reactions by means of very short pulses of light energy
1968	Lars Onsager (USA)	discovery of reciprocal relations, fundamental for the thermodynamics of irreversible processes
1969	Derek Barton (UK) and Odd Hassel (Norway)	concept and applications of conformation in chemistry
1970	Luis Federico Leloir (Argentina)	discovery of sugar nucleotides and their role in carbohydrate biosynthesis
1971	Gerhard Herzberg (Canada)	research on electronic structure and geometry of molecules, particularly free radicals
1972	Christian Anfinsen (USA), Stanford Moore (USA), and William Stein (USA)	work in amino-acid structure and biological activity of the enzyme ribonuclease
1973	Ernst Fischer (West Germany) and Geoffrey Wilkinson (UK)	work in chemistry of organometallic sandwich compounds
1974	Paul Flory (USA) macromolecules	studies of physical chemistry of
1975	John Cornforth (UK)	work in stereochemistry of enzyme-catalysed reactions
	Vladimir Prelog (Switzerland)	work in stereochemistry of organic molecules and their reactions
1976	William Lipscomb (USA)	study of structure and chemical bonding of boranes (compounds of boron and hydrogen)

Year	Winner(s)[1]	Awarded for
1977	Ilya Prigogine (Belgium)	work in thermodynamics of irreversible and dissipative processes
1978	Peter Mitchell (UK)	formulation of a theory of biological energy transfer and chemiosmotic theory
1979	Herbert Brown (USA) and Georg Wittig (West Germany)	use of boron and phosphorus compounds, respectively, in organic syntheses
1980	Paul Berg (USA)	biochemistry of nucleic acids, especially recombinant DNA
	Walter Gilbert (USA) and Frederick Sanger (UK)	base sequences in nucleic acids
1981	Kenichi Fukui (Japan) and Roald Hoffmann (USA)	theories concerning chemical reactions
1982	Aaron Klug (UK)	determination of crystallographic electron microscopy: structure of biologically important nucleic-acid–protein complexes
1983	Henry Taube (USA)	study of electron-transfer reactions in inorganic chemical reactions
1984	Bruce Merrifield (USA)	development of chemical syntheses on a solid matrix
1985	Herbert Hauptman (USA) and Jerome Karle (USA)	development of methods of determining crystal structures
1986	Dudley Herschbach (USA), Yuan Lee (USA), and John Polanyi (Canada)	development of dynamics of chemical elementary processes
1987	Donald Cram (USA), Jean-Marie Lehn (France), and Charles Pedersen (USA)	development of molecules with highly selective structure-specific interactions
1988	Johann Deisenhofer (West Germany), Robert Huber (West Germany), and Hartmut Michel (West Germany)	discovery of three-dimensional structure of the reaction centre of photosynthesis

Year	Winner(s)[1]	Awarded for
1989	Sidney Altman (USA) and Thomas Cech (USA)	discovery of catalytic function of RNA
1990	Elias James Corey (USA)	new methods of synthesizing chemical compounds
1991	Richard Ernst (Switzerland)	improvements in the technology of nuclear magnetic resonance (NMR) imaging
1992	Rudolph Marcus (USA)	theoretical discoveries relating to reduction and oxidation reactions
1993	Kary Mullis (USA)	invention of the polymerase chain reaction technique for amplifying DNA
	Michael Smith (Canada)	invention of techniques for splicing foreign genetic segments into an organism's DNA in order to modify the proteins produced
1994	George Olah (USA)	development of technique for examining hydrocarbon molecules
1995	F Sherwood Rowland (USA), Mario Molina (USA), and Paul Crutzen (Netherlands)	explaining the chemical process of the ozone layer
1996	Robert Curl Jr (USA), Harold Kroto (UK), and Richard Smalley (USA)	discovery of fullerenes
1997	John Walker (UK), Paul Boyer (USA), and Jens Skou (Denmark)	study of the enzymes involved in the production of adenosine triphospate (ATP), which acts as a store of energy in bodies called mitochondria inside cells
1998	Walter Kohn (USA), John Pople (USA)	research into quantum chemistry
1999	Ahmed Zewail (USA)	studies of the transition states of chemical reactions using femtosecond spectroscopy
2000	Alan J Heeger (USA), Alan G MacDiarmid (New Zealand), and Hideki Shirakawa (Japan)	their roles in the development of electrically conductive polymers

[1] Nationality given is the citizenship of recipient at the time award was made.

Index

Note: page numbers in *italics* refer to illustrations